T0208090

essentials

essentials provide up-to-date knowledge in a concentrated form. The essence of what matters as "state of the art" in the current professional discussion or in practice. *essentials* inform quickly, uncomplicatedly and comprehensibly

- as an introduction to a current topic from your field of expertise
- as an introduction to a subject area that is still unknown to you
- as an insight, in order to be able to speak on the subject

The books in electronic and printed form present the expert knowledge of Springer specialist authors in a compact form. They are particularly suitable for use as eBooks on tablet PCs, eBook readers and smartphones. *essentials:* Knowledge modules from economics, social sciences and the humanities, from technology and the natural sciences, as well as from medicine, psychology and the health professions. From renowned authors of all Springer publishing brands.

Klaus Stierstadt

Our Climate and the Energy Problem

How our Energy Needs can be Covered in a Climate-Friendly Way

Klaus Stierstadt
Faculty of Physics
University of Munich
Munich, Germany

ISSN 2197-6708 ISSN 2197-6716 (electronic)
essentials
ISBN 978-3-658-38312-1 ISBN 978-3-658-38313-8 (eBook)
https://doi.org/10.1007/978-3-658-38313-8

The translation was done with the help of artificial intelligence (machine translation by the service DeepL.com). A subsequent human revision was done primarily in terms of content.

This Springer imprint is published by the registered company Springer Fachmedien Wiesbaden GmbH, part of Springer Nature.
The registered company address is: Abraham-Lincoln-Str. 46, 65189 Wiesbaden, Germany

Preface

Last year, two important conferences organised by the UN took place, dealing with the future of humanity and our environment: The 6th World Population Conference in Nairobi with 6000 participants from 160 countries and the 25th World Climate Change Conference in Madrid with 25,000 participants from 200 countries. At these conferences, the problems of overpopulation and climate change were discussed in detail. But, unfortunately, at neither of them were concrete decisions taken or binding measures adopted that could positively control the developments that threaten us.

This *essential* discusses the causes that have led to climate change. It is closely related to humanity's energy needs. It is true that the sun constantly provides us with more energy than we require. However, we don't have enough of the form of energy we need, and we don't have it at all times, in all places, at a price we can bear. *And that is precisely the problem.* The reasons are largely economic.

In the first chapter, we discuss the present situation; in the second, the climate and how it comes about; in the third, the carbon dioxide problem; in the fourth, the energy requirements of a constantly growing human race; in the fifth, the properties of solar energy; and in the sixth chapter, the functions of technical energy converters necessary today. In the seventh chapter, we take a look at some non-solar or utopian energy sources.

Climate change is first and foremost an energy supply problem, and this is partly a consequence of population growth. In industrialised countries, there is generally enough energy. In the developing and emerging countries with strong population growth, however, there is too little.

München, Germany Klaus Stierstadt

What You Can Find in This *essential*

- You will gain an insight into the causes and consequences of climate change as well as an overview of the possible uses of solar energy.
- You will learn how the current global warming is caused by the production of carbon dioxide. And you will gain an insight into the problem of electricity generation from fossil fuels.
- You will learn about the properties of the sun and the devices used to convert its radiation in the form of light and wind power into electricity.
- You will learn how the growing energy demand of the increasing human population can be satisfied by using solar energy.

Contents

Introduction 1

Every day we hear and read in our media the words "climate change" and "renewable energy". Obviously, there are issues, and they are somehow related. The issues in question are indeed hotly debated by experts. But if you ask them where exactly the problem lies and how it could be solved, you sometimes get very contradictory answers. The layman wonders and does not know what to believe. There are in fact two quite different opinions among the experts.

The small group of **climate optimists** say that there is nothing wrong with our weather. The fact that it is constantly getting warmer here is a completely natural process. We are in a transition from the last ice age about 20,000 years ago to the next warm period in perhaps 80,000 years. And in the process, the temperature on Earth may rise by a total of 10°. This is caused by the changing tilt of the Earth's axis relative to its orbit around the Sun. This has been known for a long time. The temperature rises by only about one ten-thousandth of a degree per year because of this process. And the sea level rises accordingly by a tenth of a millimetre. For millions of years, living nature has been accustomed to this periodic alternation of cold and warm periods and has always behaved accordingly. In the cold season, life, plants and animals, orient themselves towards the equator, when the high latitudes of the earth are largely iced over. And in the warm period, these are then repopulated. – So much for the climate optimists. Unfortunately, they fail to mention that the current rise in temperature and sea level is occurring at least a thousand times faster than can be expected on the basis of the Earth's axis movement. So this cannot be responsible for it.

The **climate realists** are of a completely different opinion. These experts – and they are by far in the majority [5, 6] – are convinced that the current global warming of about 1° on average over the last 150 years is of human origin. It is mainly

K. Stierstadt, *Our Climate and the Energy Problem*, essentials,
https://doi.org/10.1007/978-3-658-38313-8_1

due to the increase of the gas carbon dioxide (CO_2) in the lower atmosphere, an increase from about 0.3 to 0.4 parts per thousand of the air volume. And this CO_2 has been created by the burning of vast quantities of fossil fuels – coal, oil and natural gas – over the past 150 years. Warming also leads to a higher rate of evaporation of sea water and thus to an increase in the water content of the atmosphere, which is also harmful to the climate. The increase in temperature due to the burning of fossil fuels is about a thousand times greater than that due to the tilt of the earth's axis, and this is therefore ruled out with certainty as the cause of our climate change. Sea levels are also rising at a global average rate of about one centimetre per year, because the ice reserves in glaciers, Greenland and Antarctica are melting, and also because the sea water is expanding due to warming.

This booklet discusses the real causes of climate change and their backgrounds: the increase in temperature is nothing other than a consequence of mankind's ever-growing energy consumption. And its energy needs are currently being met by the wrong means, namely with fossil fuels instead of solar energy. The growth of the earth's population by currently about 83 million per year further aggravates the problem. And this increase requires rising economic growth. This is where the cat bites the tail: with our current energy technology, economic growth means increasing carbon dioxide emissions into the atmosphere and global warming. But in no case is climate change an invention of the communists to harm the American economy, as the president of a Western state recently claimed.

2 The Climate

The earth is getting warmer all the time. We can feel this not only in our own bodies. Measurements taken in a wide variety of places all show a similar picture to that in Fig. 2.1. The average global temperature has risen by about 1° over the past 150 years, and in some places by up to 3°. Now, one might think that 1 or 2° warmer would not be much. After all, that is only a few percent of the normal seasonal variation of about ± 20° in many places. But experience teaches us something different: in Central Europe, summer periods are up to 10° hotter than they were about 60 years ago. And in winter there is hardly any snow left in our lowlands. The glaciers are melting rapidly, polar bears have to swim far and skiers have to go higher and higher. The global average of 1° hides the fact that the increase varies greatly in geography and time. It is stronger in the northern hemisphere than in the southern, and more pronounced over the continents than over the oceans. A global temperature increase of "only" 1°, for example, would be comparable to the temperature difference of ± 5° between ice ages and warm periods in the history of the Earth. The cause of the current temperature rise is everywhere the increase of the trace gas carbon dioxide (CO_2) in the troposphere.[1] This correlation has been known for decades. From 1860 to 2019, the CO_2 content has increased from 0.28 to 0.41 parts per thousand, i.e. by almost 50% (see Fig. 2.1). In addition to carbon dioxide, several other gases are involved in global warming, namely methane with about 16%, halogenated hydrocarbons with 12%, nitrous oxide with 6% and water (steam) with about 25%. Satellite measurements have been used at various locations to

[1] The troposphere is the lower part of the atmosphere in which weather events take place. In Central Europe, it reaches up to an altitude of about 12 km.

K. Stierstadt, *Our Climate and the Energy Problem*, essentials,
https://doi.org/10.1007/978-3-658-38313-8_2

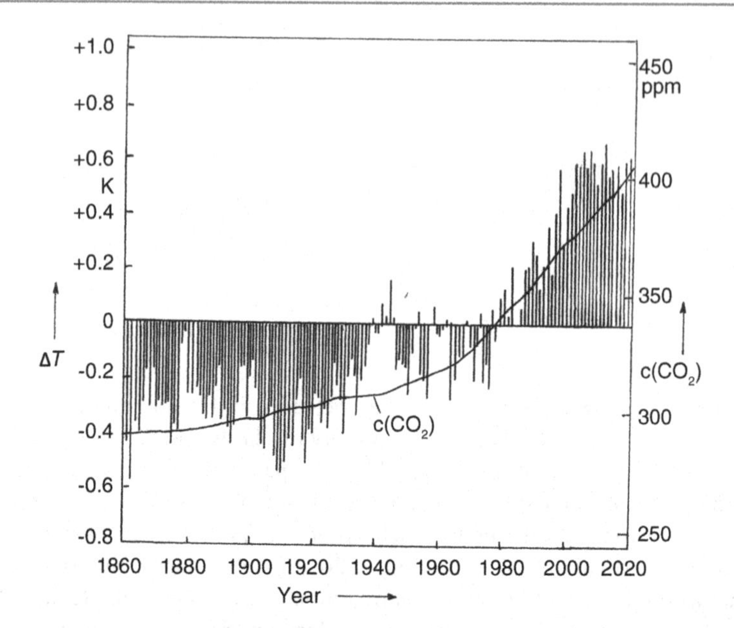

Fig. 2.1 Changes ΔT in mean global temperature at the Earth's surface (bar) and mean global CO_2 content *c of* near-surface air (line) over the past 150 years, normalized to zero in 1980. 400 ppm is a volume fraction of 400 millionths. The CO_2 content before 1960 was determined from ice cores. (After [3])

determine the CO_2 concentration and simultaneously the warming of the air near the ground [1]. The results of a model calculation are in complete agreement with this [2].

The relationship between the temperature of the troposphere and its CO_2 content is also well known from Earth history [30]. From air inclusions in Antarctic ice cores, it was possible to determine both variables quite accurately (Fig. 2.2). They run largely parallel to each other. The temperature is obtained from the ratio of heavy to light hydrogen in the ice, because this depends on the evaporation temperature. The CO_2 content can be measured directly in the ice. It increased regularly during the warm periods because of the warming of the oceans, but was always below 280 ppm [31]. It was not until 1860 that it significantly exceeded this mark (see Fig. 2.1).

We first want to understand how the relationship between CO_2 and temperature comes about. To do this, we look at the radiation budget of the atmosphere in Fig. 2.3: on a global average, the sun provides us with a radiation output (light and

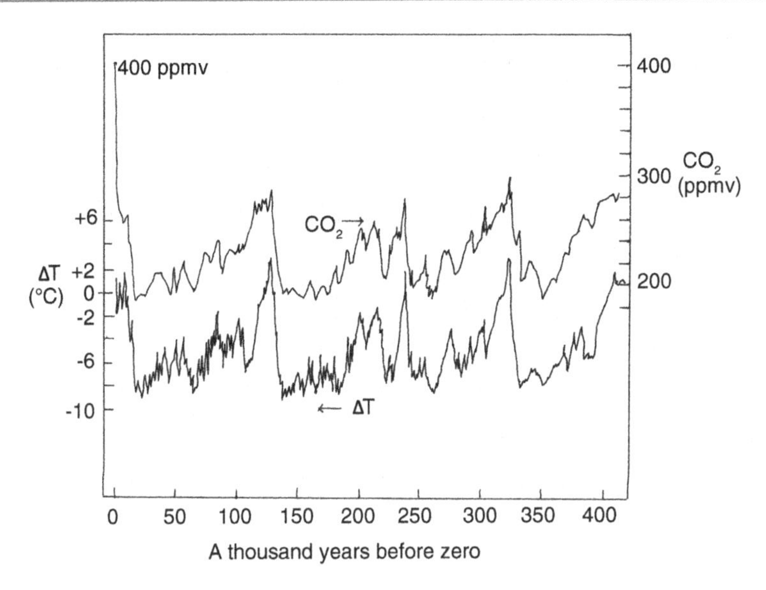

Fig. 2.2 Course of temperature and CO_2 content of the air near the ground during the past four warm and ice ages. (After [30]); the unit ppmv means millionth part by volume

heat) of 168 watts per square metre on the earth's surface. This is averaged over day and night, all latitudes, summer and winter, sunshine and rain, etc. If this radiation were completely absorbed by the troposphere and by the earth, it would of course become warmer and warmer here. But this is not the case, because the average temperature in the atmosphere near the ground has been constant for centuries at about 14 °C. This means that the energy of the incident radiation must be completely re-emitted on average so that the temperature remains stable. This can also be seen in Fig. 2.3 at the very top, because 107 + 235 is 342. But further down, within the troposphere, it gets more complicated. Various energy flows go back and forth between the air mass represented here by clouds and the Earth. The surface radiation emanating from the Earth, with a wavelength of about 15 μm in the infrared, is absorbed by the CO_2 molecules in the air, among others, and excites them to vibrate (Fig. 2.4). However, the radiation of these oscillations is not spatially oriented like the surface radiation, but is emitted isotropically in all directions. Half of it is oriented upwards and half downwards. What is emitted downwards additionally heats the earth (far right in Fig. 2.3). And this warming, apart from the water content of the atmosphere, is essentially due to atmospheric carbon dioxide and has averaged 1° since 1860. This process is called the **greenhouse effect** because the

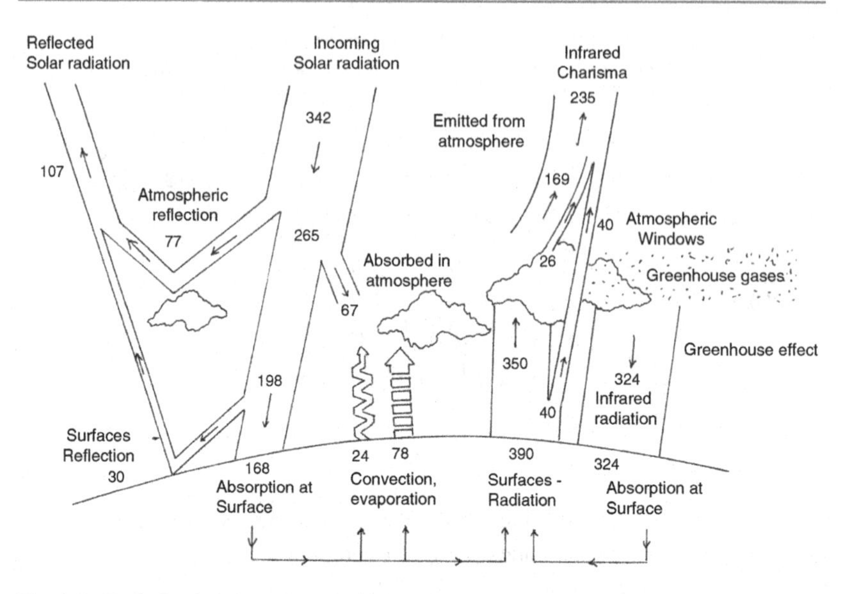

Fig. 2.3 Radiation balance of the Earth's atmosphere, averaged over the whole Earth, day and night, summer and winter. All figures in watts per square metre, explanations in the text. (After [6])

CO_2 in the air acts in a similar way to the glass windows of a greenhouse [25]. The trend in CO_2 concentration since 1860 in Fig. 2.1 suggests that the corresponding temperature increase is man-made, due to the burning of fossil fuels in the course of industrial development. In Fig. 2.3, this would correspond to an additional irradiance of the Earth's surface of about 2.5 W per square meter [6].

The observed global warming has another worrying effect: the ice is melting all over the world and therefore the sea level is rising worldwide. The glaciers and ice caps of the mountains, the Greenland ice and that of the Arctic and Antarctic contain many millions of cubic kilometres of water ice. Since the beginning of industrial evolution around 1860, sea level has risen by about 30 cm worldwide, and since 1950 this rise has accelerated (Fig. 2.5). If CO_2 production were stopped immediately today – which is, of course, illusory – the ice would continue to melt for many hundreds of years due to the CO_2 accumulated in the atmosphere, and sea level would continue to rise accordingly, initially by 30–60 cm by the year 2100. If, on the other hand, we were to continue burning all known coal, oil and gas reserves, global temperatures would rise by about 6° to 9° [6], and all ice would melt

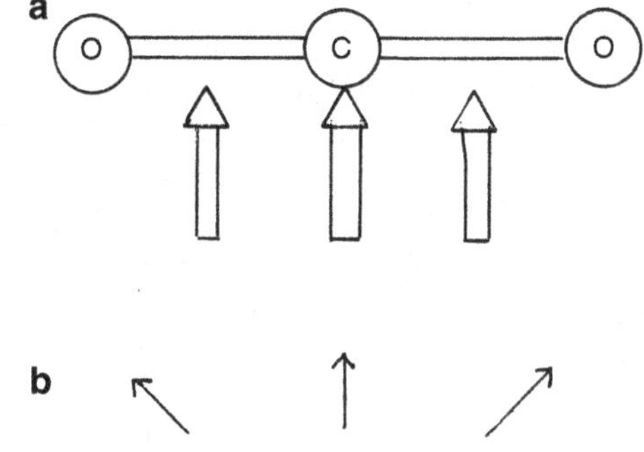

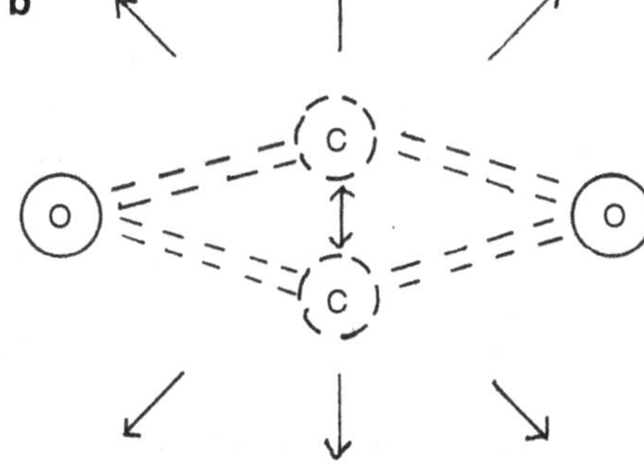

Fig. 2.4 (**a**) Structure of a CO_2 molecule and (**b**) Oscillatory motion of the same under the influence of incident infrared radiation (double arrows). The single arrows indicate the radiation emitted by the oscillation

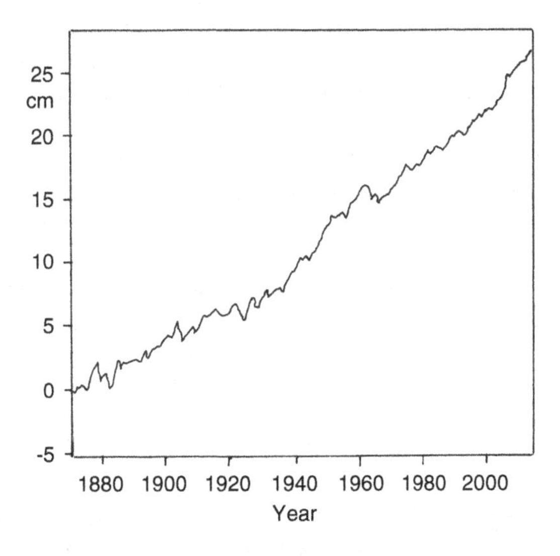

Fig. 2.5 Global sea-level rise since 1870. (After [7])

over time. Glaciers and ice caps would then yield a sea level rise of 0.5 m, the Greenland ice one of 7 m and that of Antarctica 60 m! The latter would probably not occur for about 300 years. Then we would again have conditions on earth like 35 million years ago in the Eocene or like in the last warm period 125,000 years ago: "no ice but more sea". By the way, the melting of the Arctic sea ice alone does not provide an increase, because it itself floats on the water. Anyone can easily check this at home with a drink in which an ice cube is floating: as it melts, the liquid level remains constant. Today, one third of the Arctic sea ice has already melted compared to 1950, and the North Pole has been ice-free several times in summer.

If we become sensible and stop burning fossil carbon products altogether by about 2050, then perhaps only part of the Greenland ice would have melted in the next 100 years; exactly how much is impossible to predict. This would cause sea levels to rise by 1–2 m, flooding large areas of low-lying mainland. But the people living there would then have enough time to migrate to higher-lying areas. Particularly at risk are the Pacific islands as well as Bangladesh, Egypt, Pakistan, Indonesia, Thailand, Florida and the Netherlands and the Frisian islands. In these regions, many areas will definitely have to be evacuated by the end of the twenty-first century, because large-scale inundation measures would be unaffordable. A sea level rise of just 1 m, as is expected over the next 100 years, would permanently inundate at least 150,000 km^2 of mainland and leave 180 million people homeless. The agricultural food base for part of the earth's population would be destroyed. The almost certain complete melting of Arctic sea ice in the next 50 years also has consequences: Polar bears and other polar animals would have to be relocated, to Greenland or Antarctica. Possibly the Gulf Stream would shift, and then we would have to expect a climate in Central Europe like the one we have today in Northern Norway. So we can be curious: With the Gulf Stream we'll be warmer, without it we'll be colder.

Sometimes one hears the following argument from climate optimists: our current warming is not man-made, but a natural process, because there had already been a similar warm period a thousand years ago (Fig. 2.6). Between 950 and 1100 AD, the temperature in Central Europe was at times up to 0.5° higher than before and after. This argument is wrong for the following reasons [27]: First, the temperature rise at that time affected only parts of the northern hemisphere; in the south it was colder than normal at that time. Second, the rise was much slower than today, averaging 0.2° in 100 years (1.0° today). Thirdly, the natural causes of the medieval warming are quite well known: the activity of the sun was particularly high during the period under consideration, and the irradiance of the earth was 0.4 W per square metre higher than normal. And volcanism was particularly low

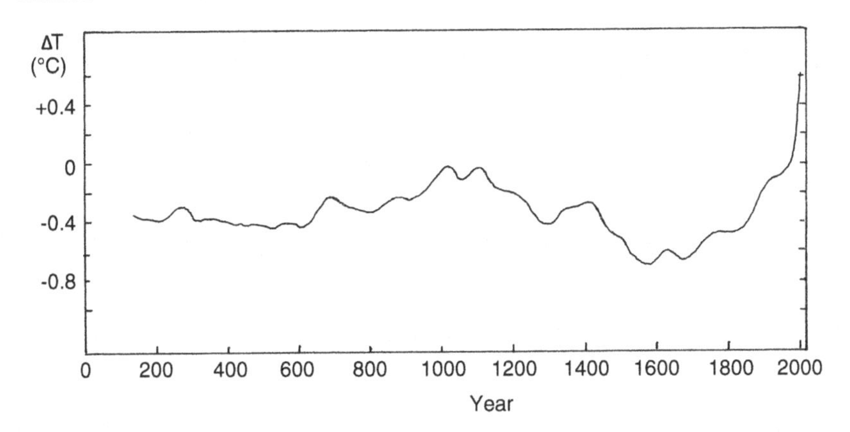

Fig. 2.6 Temperature variations of near-surface air over the past 2000 years. (After [27]). Indirect measurements before 1860, afterwards with thermometers; 1950 normalized to zero

during the same period. Fourth, the CO_2 content of the air was higher than normal during this period. This was because the warming of the sea water as a result of more solar radiation led to the release of more carbon dioxide, and likewise the burning of a great deal of wood from the increased deforestation that was taking place at that time. All of these phenomena provided additional warming. One wonders, of course, how temperatures could have been determined so accurately a thousand years ago, even though there were no thermometers? A number of indirect methods are known: Growth rings on trees, ice cores, pollen, corals, sediments, etc. All these methods yielded temperature curves similar to those in Fig. 2.6, but this does not explain the drastic increase over the last 150 years.

The conclusion of this chapter is that if we want to stabilise our climate in the long term, we must stop burning fossil carbon compounds as quickly as possible. This must happen within the next 20–30 years at the latest. In view of the dwindling reserves of fossil raw materials (see Fig. 4.5), we could then switch almost completely to solar energy, because the sun provides us with about 10,000 times our current needs free of charge (see Chap. 5).

The Carbon Dioxide CO_2

3

So how is the carbon (substance) dioxide in our air, which is changing our climate so significantly, produced? It is a colourless and odourless gas, like most other components of our air, and it is formed when the element carbon (C) burns with oxygen (O_2) according to the gross reaction

$$C + O_2 \rightarrow CO_2 + Q. \tag{3.1}$$

This is an exothermic reaction ($Q > 0$), and in the process the binding energy $Q \approx$ 4.1 eV per molecule or 394 kJ per mole is released as heat energy [8]. We use this for heating as well as in heat engines to produce motion or electricity. Electrical energy today is produced primarily by gas or steam turbines, and kinetic energy in our automobiles, ships, and airplanes by means of engines, turbines, and jet engines. The carbon-containing fuels we use today are coal, crude oil and natural gas, each accounting for about one third. This produces about 0.9 kg of carbon dioxide per kilowatt hour generated [5]. And worldwide, these are considerable quantities. Germany's total energy consumption today corresponds to 2500 terawatt hours per year,[1] or an output of about 285 gigawatts. For the whole world, the figure is around 12,000 gigawatts. At the same time, 1 billion tons or 500 km^3 of CO_2 are produced annually in Germany, and 37 billion tons or 18,000 km^3 worldwide [34]. These gigantic quantities escape from the chimneys of our houses and power plants as well as from the exhaust openings of our motor vehicles. And, as mentioned in the previous chapter, they will contribute to further warming of the air and further

[1] 1 GW is 1 million kW, 1 TW is 1 billion kW. A gigawatt, for example, corresponds to the output of a large power plant.

K. Stierstadt, *Our Climate and the Energy Problem*, essentials,
https://doi.org/10.1007/978-3-658-38313-8_3

melting of the ice. The CO_2 in the air is very stable there; it will remain for centuries if not removed or converted [5, 6].

However, if we want to stabilize our climate, we have to end this gigantic "air pollution" as soon as possible. And that is the demand of the Paris Climate Conference of 2015, namely to limit the temperature increase worldwide to 1.5° compared to the 1990 value; and this goal is to be reached by 2050. But how could this be done? In principle, it is quite simple [8]: one would have to filter out the CO_2 produced during combustion from the exhaust air and then dispose of it safely, i.e. store it or destroy it. The latter, i.e. decomposition into carbon (C) and oxygen (O_2), is of course out of the question, because this would require at least as much energy as was previously generated during combustion. So the only option is to safely store or sequester the CO_2 (from the Latin sequestere, to separate). Today, a whole range of different methods are being discussed for this purpose, but all of them are still at the experimental stage or before. The only thing they all have in common is that they cost a lot of money. It is roughly estimated that the cost of fuel, heating and electricity will increase by at least 30%. Before CO_2 can be stored, it must be separated from the combustion gases. This can be done by chemically binding the remaining components of the exhaust gas. The remaining CO_2 would then be liquefied under high pressure and could be transported in this way.

But where? And what do you do with it? The most commonly tested method would be to deposit the CO_2 deep in the ground in the hope that it will stay there. This can be done either in saline groundwater, so-called "aquifers", or in cavities created during coal, oil and natural gas extraction. There, at the high pressure corresponding to the depth, the CO_2 remains partly liquid and partly stable in a supercritical state. However, such reservoirs are never completely tight. Over time, some of the carbon dioxide diffuses out of the ground again, roughly 1% of the stored amount per year. This means that after about 300 years it will all be out in the air again, so it is not a process for eternity. One hopes that our great-great-grandchildren will then come up with something against it!

Another much discussed method would be the introduction of CO_2 into porous silicates of the earth's crust, for example also into volcanic lava. This produces exothermic carbonates and silicic acid, both stable solid compounds. However, their formation is very slow at the prevailing ambient temperatures, with a half-life of one or a few years.

A completely different and relatively cheap method of CO_2 disposal is also the afforestation of forests. This is because CO_2 is consumed during the photosynthesis of plants and converted into organic carbon compounds. However, the resulting wood must not be burned again! – And finally, it is of course possible to directly condense the CO_2 that is already in the air. However, this requires huge cooling and pressure units, and the process cost about 50 cents per kilowatt hour previously produced, about three times the current consumer price for industry.

What all the methods discussed so far have in common is that only small test facilities exist to date. It is therefore only possible to make a rough estimate of the costs that will arise if the methods are implemented on a large scale. It is also not known whether sufficient locations and storage sites exist for the quantities of CO_2 to be stored. In any case, sequestration does not come cheap. Energy prices would rise by 30 to 50% as a result. If that's too much for you, you need to seriously consider alternative methods of energy production. And as has been mentioned several times, solar energy is the cheapest and perfectly adequate alternative today.

The Energy Demand

4

Why do we produce so much carbon dioxide in the first place? Quite simply because we need so much energy and produce it predominantly with fossil fuels, with coal, oil and natural gas. Of course, we could satisfy all our needs without generating CO_2, namely from solar energy. There are mainly economic reasons why we do not do this. On the one hand, the technology for this has only been developed for a few decades, and large investments are still required. On the other hand, fossil fuels are currently still generating much greater profits than solar technology. Let's take a look at Fig. 4.1 to see what sources our energy comes from worldwide today: four-fifths of it is generated by burning coal, oil and natural gas. The use of biomass (plants, wood, algae, etc.) for food, cooking, heating, etc., also naturally releases CO_2 into the atmosphere. Solar energy itself contributes only about 2% of the total worldwide demand to date. The sources in Fig. 4.1 are collectively referred to as **primary energy.** On the other hand, let us now look at the "sinks" of energy in Fig. 4.2, i.e. what is available for the various types of consumption, this time for Germany alone. Here we see that trade, transport and industry together consume three quarters of the available energy, mainly in the form of heat, chemical energy and electricity. All these sinks are called **secondary energy** or **final energy.**

- Here is a comment on the use of language: We always talk about "generation" or "consumption" of energy. This is physically wrong. Energy cannot be produced or consumed. Its quantity remains constant in every process. What changes is only the *type* or *form of energy.* This is a law of nature. Such forms are heat, light, motion, electrical or magnetic energy, mass energy, potential energy, etc. When one of these forms is "consumed," that is, decreases, the same amount of energy is simultaneously "created" in another form. For example, in

K. Stierstadt, *Our Climate and the Energy Problem*, essentials,
https://doi.org/10.1007/978-3-658-38313-8_4

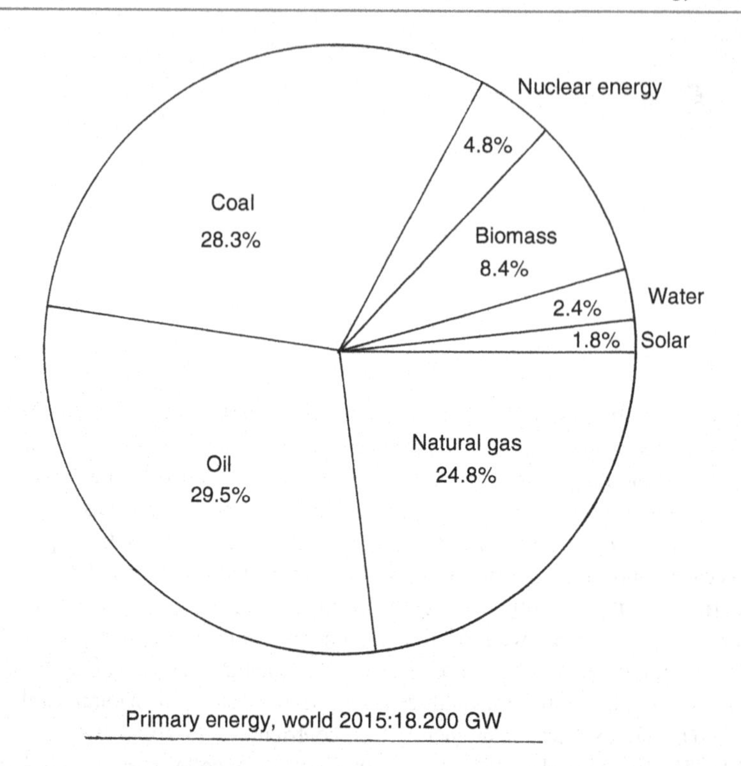

Fig. 4.1 Shares of the various primary energy sources in the world's total output in 2015, which amounted to 18,200 GW. If this is extrapolated to 2020 with 2% growth, this results in 20,020 GW per year [9]. Germany alone accounted for 440 GW of primary energy in 2015

a car engine, chemical energy is converted into mechanical motion and heat, or in a power plant, mechanical energy is converted into electrical energy, etc. But nothing is lost in total.

What can we learn from Figs. 4.1 and 4.2? First and foremost, it is striking that consumers, for example in Germany, receive about 35% less energy than is taken from the sources (290 instead of 440 gigawatts). Where has this difference gone? Well, it has gone into the environment as useless heat at the power plants, into the cooling water and into the air via the cooling towers. This is largely a consequence of the *second law of thermodynamics* [16]. It is a law of nature and requires that the physical quantity *entropy* remains constant or increases in any process in a closed system. And entropy, roughly speaking, is, for example, the ratio of exchanged heat

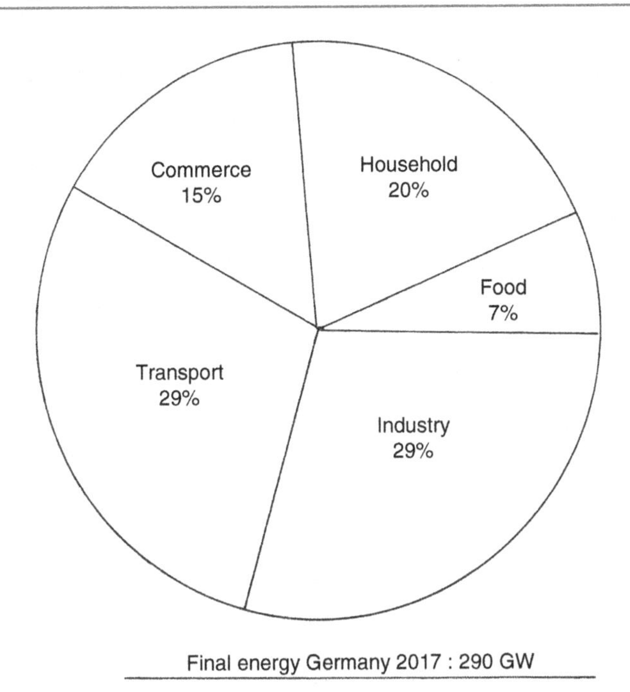

Fig. 4.2 Shares of secondary energy ("final energy") required in Germany for various sectors in 2017 as a proportion of total output. The distribution among the various consumption sectors corresponds to the usual ratios in highly industrialised countries [10, 26]

energy Q to temperature T. Therefore, part of the heat generated during combustion must always flow into the colder environment so that Q/T increases, and globally this is the missing 35% of the energy. However, clever engineers could reduce this proportion. To do so, the efficiencies of our energy converters would have to be improved (see Chap. 6).

But that is not all. Figure 4.3 shows which parts of the primary and available final energy are actually used by the consumer, for example as movement, light, heating, cooling, etc. In this conversion of final energy into **useful energy,** there are again about 30% losses, which flow into the environment as heat. However, this is not the fault of the second law, but of technology or human reason. A car engine or a lamp becomes warm although mostly only movement or light should be produced. This heat is not used, and that is the missing 32%. The *gigantic waste of energy* shown in Fig. 4.3, amounting to a total of two-thirds of the effort, is crying out for improvement. A great deal can be done here, for example the following: The

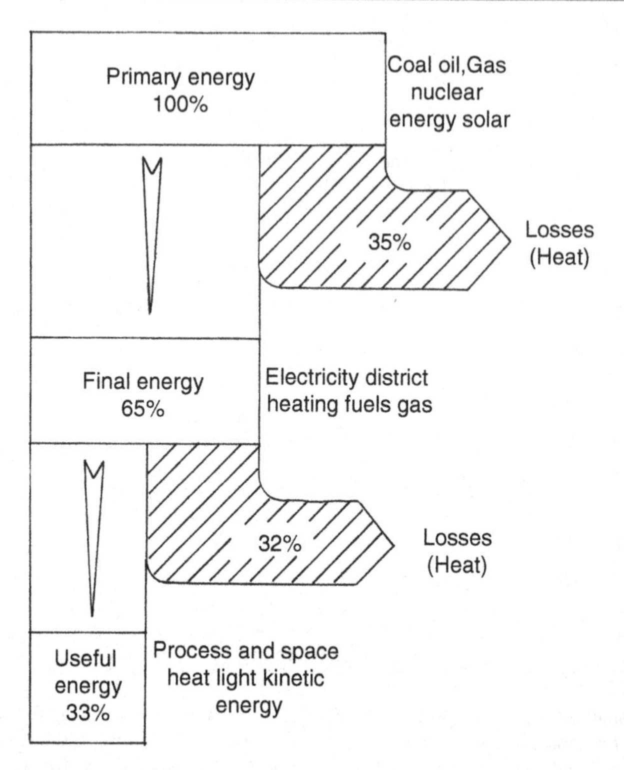

Fig. 4.3 Conversion of primary energy via final energy (secondary energy) into useful energy in Germany in 2000. Total losses amounted to 67%, i.e. two thirds (!) of the energy used. (After D. Hein, lecture 2001)

losses in the conversion of primary to final energy, especially in our power plants, could be reduced by using the waste heat as completely as possible for heating, so-called **combined heat and power,** keyword district heating. The losses in the transition from final to useful energy could be reduced by various measures: by better heat insulation of the houses, because today half of the energy used for heating is chased out through the windows and through the walls into the open air. Restricting the speed of road traffic and shifting freight transport from road to rail would also bring considerable savings. Overall, these measures could reduce total losses from 67% today to less than half. But this is opposed by massive economic interests. Industry wants to sell as many cars, fuels and heating oil as possible; the building industry wants to build as many homes and offices as cheaply as possible and the electricity companies want to sell as much electricity as possible. At least that is what the advertisements tell us day after day. Only decisive political action would bring about a change here.

And as already mentioned, the use of solar energy would largely free us from the CO_2-generating combustion of fossil raw materials (see Chap. 6). Only for air travel has a viable alternative not yet been found. But solar technology is a relatively young art, and its comprehensive use still requires large investments that will only yield respectable profits in decades. Therefore, the conversion to solar energy is still proceeding far too slowly today.

Of course, the question is how much energy or power do we really need now and in the near future? If we divide the approximately 20,000 GW calculated for 2020 in Fig. 4.1 by the number of the earth's inhabitants today, about 8 billion, we get about 2.5 kW per person. But this energy is distributed very unevenly [33]. In the industrialized countries, each person today has an average of 5.5 kW of final energy available (see Fig. 4.3), in the USA even 10.0 kW. This is the work equivalent of about 20 human workers! In the developing and newly industrializing countries, on the other hand, it is only 0.5 kW (India 0.7 kW), i.e. the energy consumption of a single physically active person. This unequal distribution leads to the strong social tensions that exist today between population groups. Most people have too little energy or it is too expensive for them. This cannot go well in the long run.

Added to this is the fact that the earth's population is constantly growing rapidly, today by about 163 per minute or 86 million per year, or by 1.15%. If we extrapolate the current growth to the year 2050, there will then be around 10 billion people on earth instead of today's 7.75 billion. The lower curve in Fig. 4.4 shows how consumption will then develop. If the uneven distribution remains, we will need about 16,000 GW as final energy in 2050 instead of today's 13,000 (the 20,020 from Fig. 4.1 minus 35% from Fig. 4.3). That is an increase of 20%. Of course, in order to protect our climate, we must under no circumstances obtain this additional demand from the burning of fossil raw materials. If we also become reasonable and even out the differences between the populations of industrialized and developing countries somewhat, we obtain the upper curve in Fig. 4.4, assuming that by 2050 the individual power demand in industrialized countries has been gradually reduced from 5.5 to 4 kW by austerity measures. In the developing countries, on the other hand, it increases from 0.5 to 2 kW. This results in a power demand of 25,000 GW in 2050, almost twice as much as today! Where will this huge amount come from? In view of the enormous potential of solar energy of around 17 million GW (see Fig. 5.1), there is probably no alternative but to use it. But if the balancing of the energy demand between the population groups seems to be a social utopia, let us wait until the poor take by force what the rich have too much of.

Using solar energy is also really the only way we have, because fossil fuel supplies are diminishing rapidly. Figure 4.5 shows how long the known quantities will

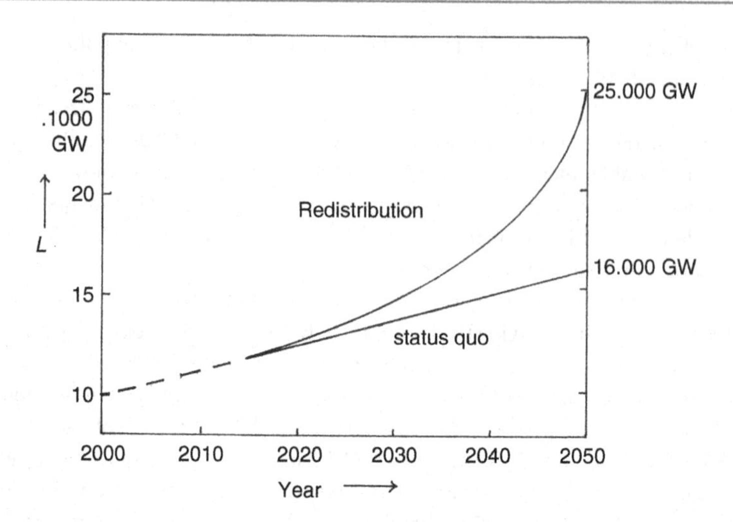

Fig. 4.4 Development of global power demand *(L)* for final energy up to the year 2050. Lower curve: Industrialized countries 5.5 kW per person and developing and emerging countries 0.5 kW, respectively. Upper curve: 4 and 2 kW, respectively. Population growth 1%/year. Population share in industrialized countries 2015 20%, 2050 25%

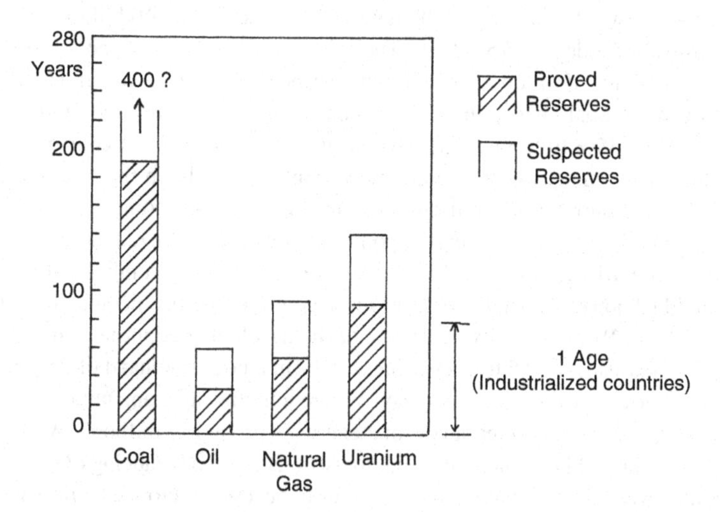

Fig. 4.5 Expected useful life of fossil fuel reserves and uranium. (After D. Hein, lecture 2001)

last at *today's consumption rates.* That's only about a generation away for oil and gas, maybe a generation and a half for uranium. Coal may last another 200 years if it is not hydrogenated on a large scale as a substitute for oil and gas [11]. However, if the growing demand shown in Fig. 4.4 is also met from existing supplies, then these will be exhausted even more quickly. One sometimes hears the opinion that there are still unknown large quantities of such raw materials in the Arctic and Antarctic. This may be true, but it should be borne in mind that the extraction and transport of such fuels will be much more expensive there than is the case today with conventional deposits, certainly much more expensive than the use of solar energy. And in that case our climate would deteriorate much faster than it does today.

5 The Sun

So far we have repeatedly referred to the practically inexhaustible supplies of solar energy. Now we want to see how this energy is created and what it does on earth [11].

Our Sun provides us with several thousand times as much energy as we need (Fig. 5.1), and will probably continue to do so for several billion years. The sun is a typical star, of which there are about 100 billion in our galaxy, the Milky Way. Its diameter (1.4 million km) is about a hundred times that of the Earth (12,740 km). Its mass ($2*10^{27}$ t) is about 300,000 times that of the Earth ($6*10^{21}$ t). The Sun is composed of about 70% hydrogen, 28% helium, and 2% other heavier elements. Figure 5.2 shows a schematic section through the Sun. At its core it is about 10 million degrees hot, at its surface only 5500 °C. The Sun, and we with it, are located in a spiral arm of our galaxy, about 28,000 light years away from its center, which it orbits once in 20 million years at a speed of 250 km per second.

The Sun radiates energy with a power of 400 million times billions of gigawatts, which we perceive mainly as light and heat. This corresponds to the power of $4\ 10^{17}$ large earthly power stations! Where does this huge energy come from? It comes from the fusion of hydrogen into helium, which takes place in the interior of the Sun. Such a process requires the high temperature of 10 million degrees prevailing there. It came into being when, during the formation of the Sun from the hydrogen present in space, its gravitational energy was converted into the kinetic energy of the atoms. At 10 million degrees, all atoms are completely ionized, that is, matter consists of a gaseous plasma of atomic nuclei and electrons. When two hydrogen atom nuclei or protons (H^+) collide in it, they form a deuterium nucleus (D^+), as sketched in Fig. 5.3. If another hydrogen nucleus is added, a helium-3 nucleus (${}_{2}^{3}He^{++}$) is formed, and two such nuclei form a stable helium-4 nucleus (${}_{2}^{4}He^{++}$),

K. Stierstadt, *Our Climate and the Energy Problem*, essentials,
https://doi.org/10.1007/978-3-658-38313-8_5

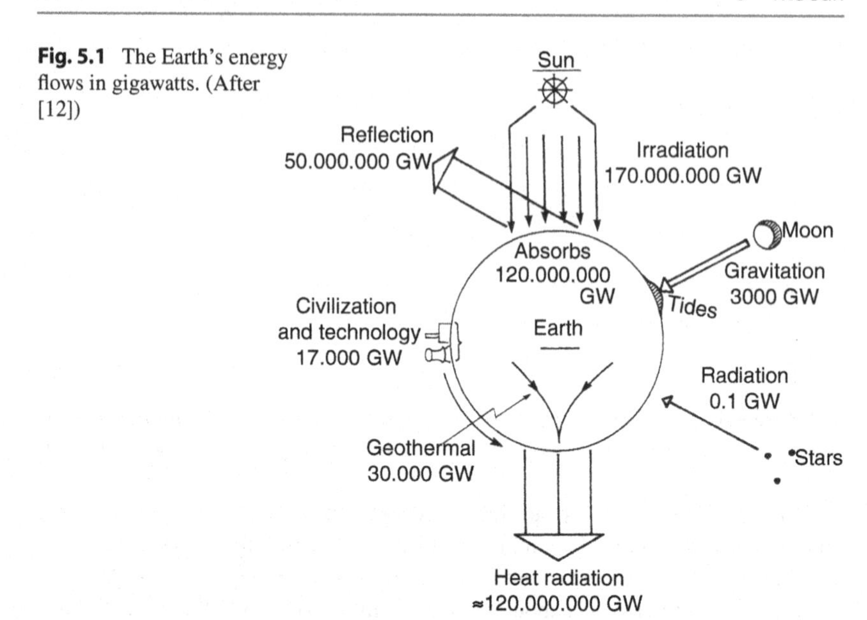

Fig. 5.1 The Earth's energy flows in gigawatts. (After [12])

again emitting two protons. In this way the hydrogen of the Sun is gradually "burned" to helium, 560 million tons per second, a cube of 80 m edge length! But because of its huge size, the Sun's hydrogen supply will last for another 6–7 billion years. So we need not fear that it will soon stop radiating.

The energy released during this fusion process initially consists of gamma radiation (γ), as outlined in Fig. 5.3. These are electromagnetic waves, similar to X-rays, but with hundreds of times higher energy or shorter wavelength. These rays diffuse from the interior of the Sun to the surface and gradually lose energy by interacting with atomic nuclei. In the process, they get longer and longer wavelengths and are finally emitted from the surface into space as visible light and as thermal radiation. Such a gamma quantum takes about 200,000 years to travel from the center to the surface. So what we see today was produced inside the Sun around the time of the Neanderthals! The light essentially comes from the photosphere of the Sun (see Fig. 5.2), whose temperature is about 5800 °C. It then spreads out in space in a straight line, and a tiny little part of it also reaches us on Earth after about 8 min. There it awaits a fate such as we had seen in Fig. 2.3. The radiation power hitting the high atmosphere is 1370 watts per square meter at perpendicular incidence of light, the so-called **solar constant.** Integrated over the entire surface of the earth, this results in the 170 million gigawatts shown in Fig. 5.1, which is about

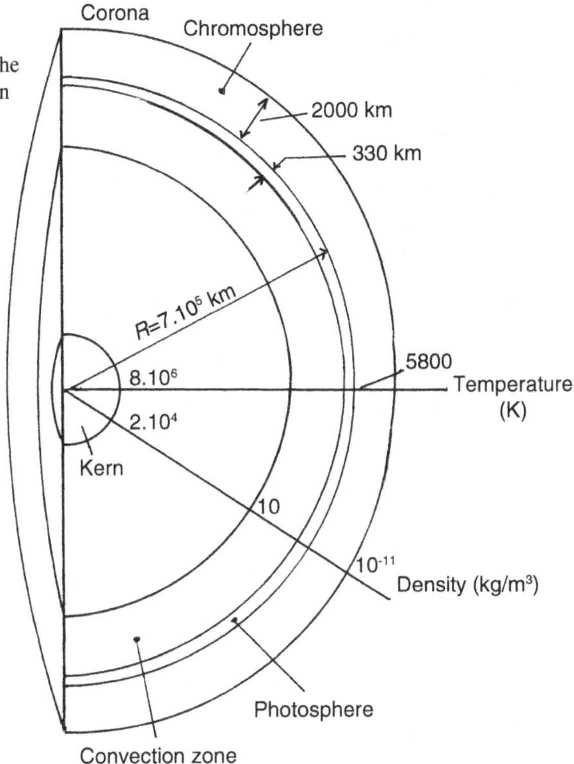

Fig. 5.2 Structure and physical data of the Sun (After [13]). In this scale, the Earth would be a pinhead in 10 m distance

8000 times the amount of energy needed by mankind today (see Figs. 4.2 and 5.1)! This clearly shows that our energy problem is not one of quantity but of quality and distribution. It depends on where and in which form we need the energy: If you lie in the Sun, you get warm but not full. And if you let the Sun shine on your car, the car does not start moving, it just heats up. So we need devices that convert solar energy into other forms of energy (electrical, chemical, kinetic, etc.) to satisfy our needs, and we discuss that in Chap. 6.

The **spectrum of Sunlight** is shown in Fig. 5.4. It shows the intensity of the radiation in watts per square metre and per wavelength interval as a function of wavelength (lower scale) or energy (upper scale). Indeed, light can be considered as a wave or as a particle (quantum or photon). The smaller the wavelength, the greater the energy. The intensity is greatest in the visible range, about ten times smaller in the near infrared, and about a hundred times smaller in the ultraviolet.

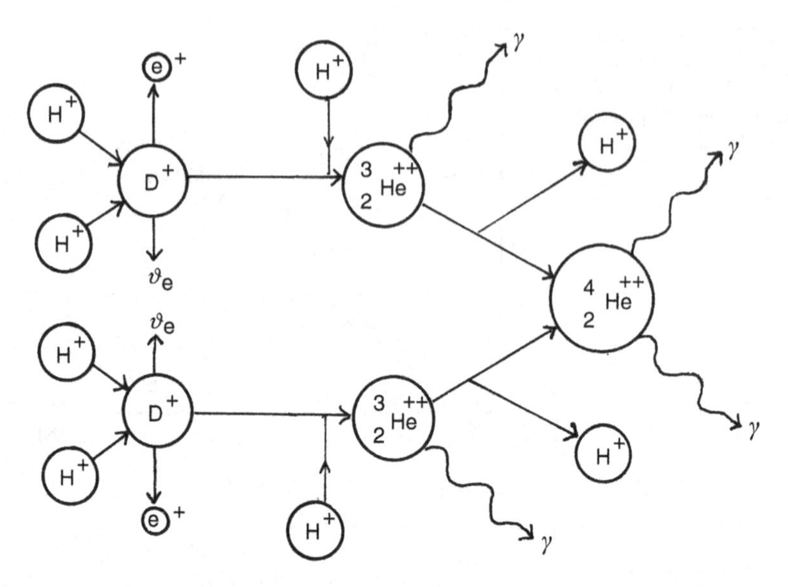

Fig. 5.3 On energy production in the Sun (e+ positron, γ gamma quantum). Explanations in the text

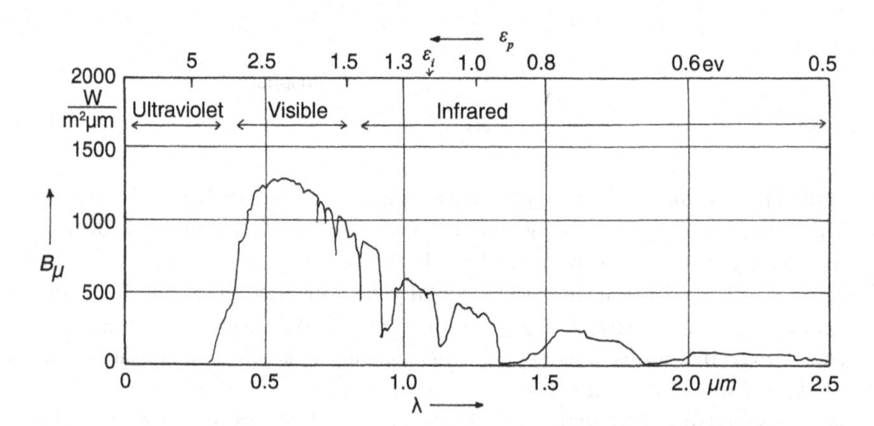

Fig. 5.4 Solar spectrum at the Earth's surface under clear skies and with the Sun at 42° above the horizon. Irradiance $B\mu$ as a function of wavelength and photon energy. The cuts in the spectrum are due to the absorption of radiation by atmospheric gases. (After [14])

The heat we feel when exposed to the Sun is caused mainly by the long-wave infrared part of the light, called thermal radiation. Ultraviolet light is used by certain plants and animals; in humans, however, it causes skin cancer.

The 170 million gigawatts that the Sun constantly supplies (see Fig. 5.1) or the 1370 watts per square metre (solar constant) are of course not completely available to us. If one averages over the whole earth, over day and night, summer and winter, and subtracts the power reflected and absorbed by the atmosphere, then around 170 watts per square metre remain at the earth's surface (see Fig. 2.3). This value still varies depending on cloud cover, dust content and temperature distribution in the troposphere. In nature and technology, the 170 watts are converted in many ways into other forms of energy, for example into motion as wind, waves, water currents, into chemical energy in plants and animals, into electricity through photovoltaics, etc. In Fig. 5.5 the earthly fate of solar radiation is explained quantitatively. Nuclear energy, geothermal energy and tidal energy, which do not come from the Sun, are also included here (see Chap. 7). Less than one per thousand of the incident radiation is converted into oxygen and biomass by the chlorophyll of green plants and algae with the help of carbon dioxide and water. This small fraction of solar energy, about 150,000 gigawatts or ten times the technical demand, is far enough to supply oxygen and food to all mankind and all living things. Biomass is used in various ways: About 700 GW is used for human food and animal feed. That is 1.5% of the total photosynthetic output. About three times that, 2000 GW, is burned in cooking and heating. And 1500 GW is used to generate electricity and produce biofuels for internal combustion engines. And finally, about 70% of the incident energy is emitted back into space as heat or infrared radiation (see also Fig. 2.3).

As mentioned at the beginning, solar energy is practically inexhaustible. The Sun was formed about 5 billion years ago from a compression of interstellar hydrogen gas. And this probably stemmed from the supernova explosion of another star. In such an event, almost all the energy of a star is released within hours and radiated into space. The stability of the Sun currently relies on a balance between fusion pressure (from within) and gravitational pressure (from without). This works well as long as there is enough hydrogen for fusion. But because this is gradually "burned" to helium, and this to carbon and other heavier elements, the fusion pressure decreases over time. Gravitational pressure then squeezes the Sun, making it hotter and hotter inside. In about 7 billion years, this heat takes over and the Sun expands to a thousand times its current radius. It becomes a red giant star, like several we know in our galaxy. Its diameter will then be ten times greater than the current radius of the Earth's orbit. All inner planets up to Mars and also the Earth are swallowed up by the Sun. A few million years later the Sun will shrink again to one tenth to one hundredth of its original size and become a white or black dwarf

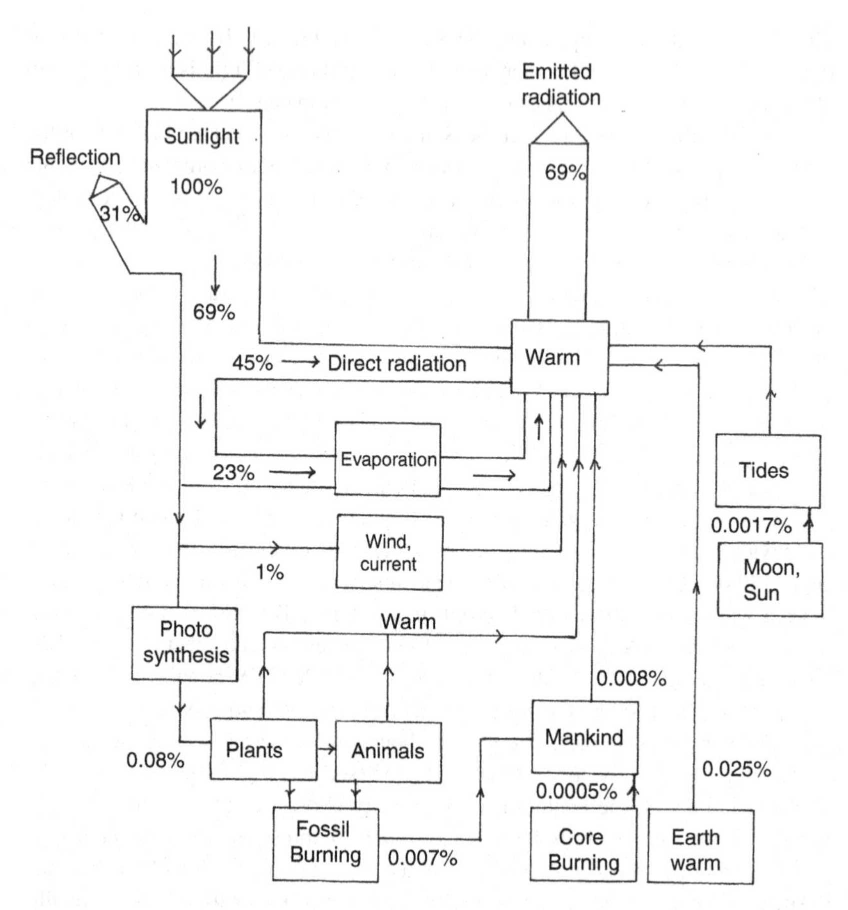

Fig. 5.5 Energy flow diagram of the Earth (After [6]). The percentages are based on slightly different calculations than in Figs. 2.3 and 5.1

star. It then shines only very faintly with one hundredth of today's intensity. But how do we know all this? One concludes it from the observation of other Sun-like stars in our galaxy, which are in the most different development stages.

So our future destiny will be: first fire, then ice. But that will happen, as I said, only in about 7 billion years. Whether human life will still exist then is rather improbable: Either we eliminate ourselves beforehand by wanton destruction of our environment or our biological basis of life [15]. Or all organic life on Earth will be wiped out by a nearby supernova explosion. The probability of this happening within 7 billion years is 100% by order of magnitude. The radiation of such an explosion consists of neutrons, protons, electrons, gamma quanta etc. and is absolutely deadly for all organic life. What remains of the exploded star is usually a black hole, of which nothing can be seen, because nothing can come out of it, not even light.

6 The Solar Energy Converters

6.1 Overview

Solar energy, which comes to us as light and heat, can be converted into other forms of energy in many ways. Humans discovered this about a million years ago when they learned to control fire. They could use it to heat and prepare cooked food. Then, about 10,000 years ago, they learned to harness the movement of wind and water, all indirect forms of solar energy. There were the first sailing ships, windmills and watermills. Today, we have a large number of technically sophisticated methods for harnessing solar energy. An overview is shown in Fig. 6.1a–c. Please delve into it a little, and you will see that the most frequently generated secondary form of energy is electrical, here referred to as "electricity" for short. There is a deeper reason for this: electrical energy is the easiest to transform into many other needed forms of energy, light, heat, cooling, motion, chemical energy, vibration energy, magnetic energy, etc.

In this chapter we want to briefly discuss the most important devices and machines that have been invented to convert solar energy, the so-called **energy converters.** The variety of such possibilities will allow us in the future to completely abandon the burning of fossil fuels and thus save our climate.

Here is a word about the use of language: The term "renewable energy" has become part of our colloquial language. This by no means means that energy can age and must be renewed, such as a living being, a building or an appliance. Rather, it refers to those forms of energy that are not derived from fossil fuels such as coal, oil, natural gas, or uranium, but from solar energy, geothermal energy, tides, etc.

K. Stierstadt, *Our Climate and the Energy Problem*, essentials,
https://doi.org/10.1007/978-3-658-38313-8_6

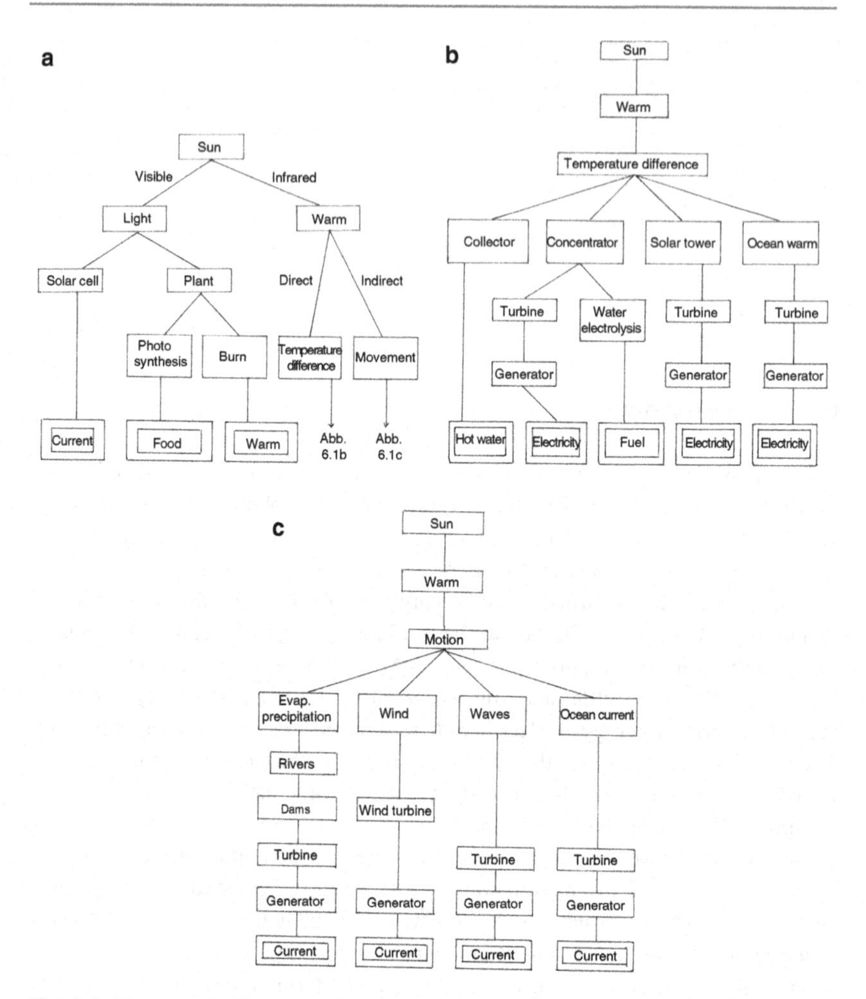

Fig. 6.1 The use of solar energy in nature and technology. The useful energy forms are outlined twice. (**a**) primary processes, (**b**) direct use via a temperature difference, (**c**) indirect use via the movement of wind and water.

The word "renewable" is out of place here. It would be better to speak of "alternative" or "inexhaustible" energy. Equally out of place is the term "sustainable" for these forms of energy. It comes from a mistranslation of the English "sustainable", which means "compatible" (for the environment).

6.2 Efficiency

A general characteristic of all energy converters is their **efficiency** η. It indicates "how good" they are or, more precisely, which part of the energy E_{zu} supplied to the converter is available again as E_{ab} dissipated:

$$\eta = \frac{E_{\mathrm{ab}}}{E_{\mathrm{zu}}}. \tag{6.1}$$

This proportion is always smaller than 1 or than 100%. The causes for the loss characterized by this are on the one hand the second law of thermodynamics [16] and on the other hand the technical imperfection of the devices and machines. Electric machines and water turbines operate with particularly high efficiencies, i.e. particularly loss-free. On the other hand, there are particularly high losses in lamps and in heat-power machines such as steam engines, diesel and petrol engines, etc. An overview of the efficiencies of the energy converters can be found in the following Table 6.1.

6.3 Solar cells

We begin our explanations with the well-known photovoltaic **solar cells** or **solar modules,** which are located on our roofs and also freely in the landscape. They convert sunlight directly into electricity (Fig. 6.1a). The structure of such a cell is sketched in Fig. 6.2. It consists of two superimposed semiconductor layers, for example of silicon. The upper layer is n-doped, i.e. it has an excess of electrons. The lower layer is p-doped and therefore has too few electrons. Some of them then migrate from the n- to the p-layer. This creates a space charge layer (RL) at their contact surface, in which a strong electric field $\boldsymbol{E}$ prevails. If a light quantum with sufficiently high energy ε enters this layer, it can ionize a silicon atom. The resulting electron is attracted upwards into the n-doped layer by the electric field. This creates an electrical voltage between the top and bottom of the cell. If the two are connected around the outside by conductors (K,K) as in the figure, an electric current flows through them. For a 1 dm^2 cell, a current of about 1 ampere is obtained when irradiated with 10 watts of light power. The modules of 1 m^2 surface that are available today contain 100 such cells and deliver an electric current of 100 W at the maximum possible light output of 1000 W in our latitudes. According to Equation (6.1), this corresponds to an efficiency of 10%. Nine tenths of the radiated energy is dissipated into the environment as useless heat. In laboratory tests,

Table 6.1 Real efficiencies of various energy converters (mean values). The conversion types are abbreviated as follows: *c* chemical, *e* electrical, *m* mechanical, *s* radiation, *t* thermal

Devices	η (%)	c → e	c → m	c → t	e → c	e → m	e → s	e → t	m → e	s → e	s → t	t → e	t → m
Electr. generator	100								●				
Electric motor						●							
Electric heating								●					
Ocean current-KW									●				
Hydroelectric power plant	90								●				
Tidal power plant									●				
Dry battery		●											
Gas heating	80			●									
Vacuum collector											●		
Power-heat coupling												●	
Accumulator	70	●											
Oil heating				●									
Combined cycle power plant												●	
Fuel cell		●											
Solar collector	60										●		
Jet engine				●									
Wind power plant									●				
Wave power plant	50								●				
Steam turbine													●
Geothermal power												●	
Diesel engine	40		●										
Gasturbine			●										
Solid state laser							●						
Solar therm. power plant											●		
Solar cell	30									●			
Gasoline engine			●										
Parabolic concentrate											●		
Fluorescent tube	20						●						
Steam locomotive			●										
Thermopile										●			
Incandescent lamp	10						●						
Updraft power plant										●			
Marine thermal power station	0											●	

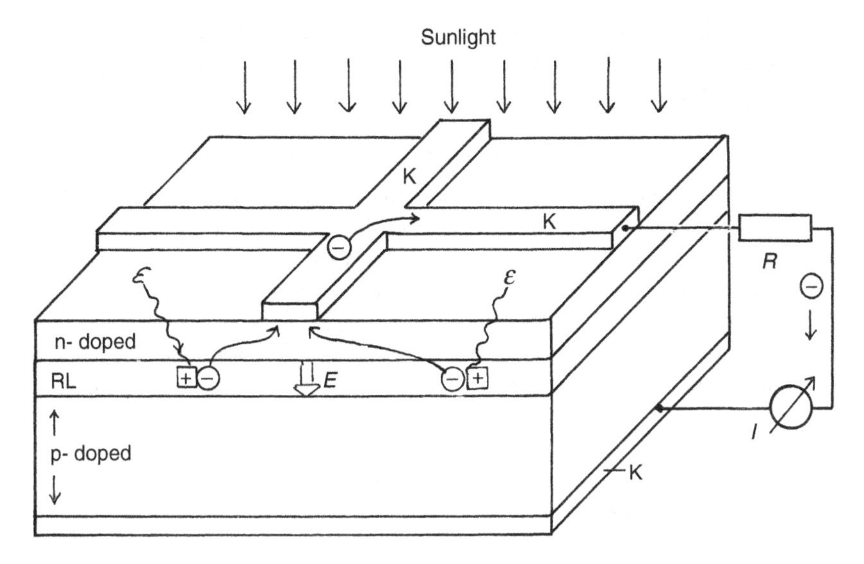

Fig. 6.2 Principle of a photovoltaic solar cell, explanations in the text

however, efficiencies of up to 40% have been achieved with specially constructed cells. In order to cover the electricity demand of approx. 1.5 kilowatts of an inhabitant of the industrialized countries, about 90 m^2 of normal module surface is required at an average irradiation of 170 watts per square meter (see Fig. 2.3). For inhabitants of developing countries with a consumption of 160 watts, however, this is only about 10 m^2.

6.4 Biomass

Next, we discuss the nature-supplied energy converter that exists on our planet, the green plants and algae. According to Fig. 5.5, they convert about one thousandth of the incident solar energy. This is about ten times the global human consumption. The plants and algae do this with the help of **photosynthesis.** This uses visible light of energy ε to produce hydrocarbons and oxygen from carbon dioxide and water. Such hydrocarbons are for example starch or glucose:

$$6\,CO_2 + 6\,H_2O + \varepsilon \rightarrow C_6H_{12}O_6 + 6\,O_2. \tag{6.2}$$

This gross reaction proceeds in plants with the help of chlorophyll molecules through a series of about 20 partial reactions [11]. All plant materials are also formed in a similar way, wood, straw, leaves, fruits, etc. The efficiency of photosynthesis is only about 0.5%, 99.5% of light energy is converted to heat. This is a typical efficiency for many organic processes.

So much for the energy conversions from the visible part of the solar spectrum. Now we come to the infrared part of the same, the "solar heat" (Fig. 6.1b,c). It can be used in two different ways: Either directly due to a resulting temperature difference, or indirectly via movements in air and water generated by it. In most conversion methods used today, the end product is electricity, or "current" for short. This is produced by a generator driven by a heat-power machine or turbine. We assume here that one knows from school how these devices are constructed and function [11].

6.5 Solar Collectors

We start with the simplest of such thermal energy converters, a **solar collector.** It consists of an arrangement of parallel tubes in which a liquid is heated by heat radiation, for example oil or water. If the collector is mounted on a sloping roof, the heated liquid rises due to its lower density and reaches the interior of the building to be heated through a system of tubes. Often the liquid is also circulated by a pump. The efficiency of such a collector is between 60 and 75%. For a four-person household in Germany, about 5 m^2 of collector area is needed to supply hot water, and about 15 m^2 for space heating. In our country, however, this is only enough to cover a quarter of the annual heating requirement.

6.6 Solar Power Plants (Solar Ovens)

A solar collector is quite effective, but it only provides relatively small temperature differences, up to about 100°. For many purposes, much higher temperatures are needed, especially for heat-power machines to produce electricity. For this purpose, one uses devices that concentrate the heat of the sun, such as the familiar burning glass or concave mirror. Figure 6.3 sketches three different such **concentrators**: a parabolic mirror, an array of plane mirrors with a central receiver, and a parabolically curved trough. With the first two, the irradiance (power per area) of the sun can be increased a thousandfold; with the parabolic trough, it can be increased a hundredfold. Figure 6.4 shows pictures of such systems. Temperatures of up to 3000 °C can be reached in the focal spots of the mirrors. This can be used to

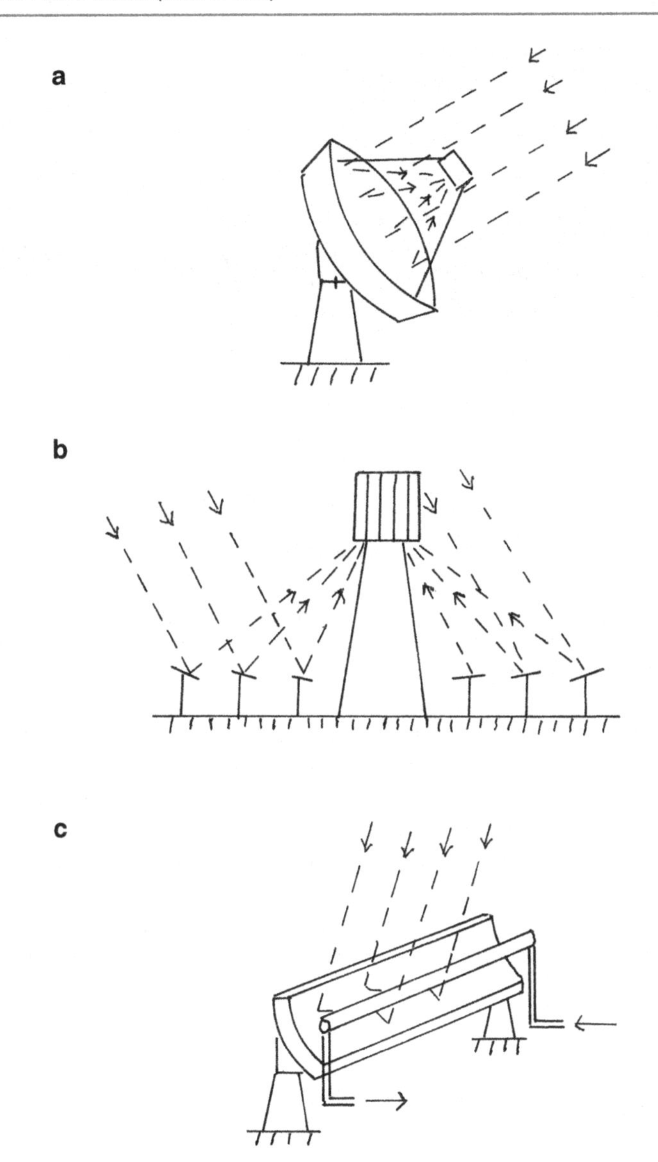

Fig. 6.3 Three designs of mirror concentrators. (**a**) Disc concentrator, (**b**) Central receiver system, (**c**) Parabolic trough concentrator

Fig. 6.4 Solar power plants. (**a**) Disc concentrator in Almeria, Spain (Photo: Lumos 3), (**b**) Central receiver system in California (Photo: abc 123), Parabolic trough concentrator in Spain (Photo: DLR/Ernsting)

heat liquids such as water, oils or molten salts, which then drive a turbine and a generator. With a parabolic mirror or disk concentrator, one thus obtains several megawatts of electrical power, about as much as a large wind turbine (see Fig. 6.5). A central receiver delivers as much as 25–50 megawatts. And a large parabolic trough plant in California's Mojave Desert today generates 350 megawatts with a mirror surface of 2.3 km^2, as much as a medium-sized coal- or gas-fired power plant. The producer price of electricity from these plants today is still about twice that of conventional power plants. However, it is expected that solar power plants will be fully competitive in about 10 years. Their efficiency for generating electricity is currently 25–30%.

6.7 Wind Turbines

Besides photovoltaics, **wind power** is the most promising supplier of solar energy today. Windmills existed several thousand years ago, but modern wind turbines only came into use after the oil crisis around 1970. Today they are up to 200 m high and the rotor blades are up to 100 m long (Fig. 6.5). About 1% of the incident solar energy (see Fig. 5.5), namely 1.7 million gigawatts, is converted into kinetic energy

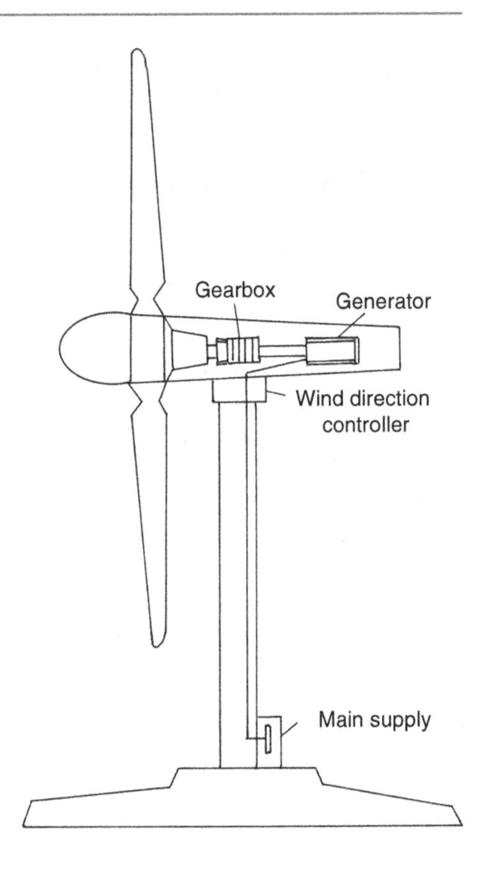

Fig. 6.5 Construction of a wind power plant

of the air worldwide (see Fig. 5.5). It is estimated that about 1% of this, or 17,000 gigawatts, could be used technically if wind turbines were installed wherever it is worthwhile and where they do not interfere. This could cover the entire energy demand of mankind in 2050 (see Fig. 4.4). Wind is therefore a very promising energy source. However, it is still used far too little, mainly for economic reasons, but also because of some reservations among the population about wind turbines. In Germany, there are currently around 25,000 of them in operation with a combined maximum output of 75 gigawatts. This amounts to 3–5 megawatts per wind turbine, but can only be used about 30% of the time due to fluctuating wind speeds. Only at speeds of 15–30 m per second (wind force 6–10) does such a wheel deliver its full power. At lower speeds it stops and at higher speeds it breaks down. The efficiency of today's wind turbines is around 0.5 or 50%. Worldwide, about 100,000 wind turbines with a maximum capacity of 400 gigawatts have been installed so far. The electricity producer price of 6 cents per kilowatt hour is already competitive with that from conventional power plants (see Fig. A1).

6.8 Efficiencies and Potentials

An overview of the *real* efficiencies of the energy converters discussed and some others was already shown in Table 6.1. It must be taken into account that the efficiencies of the individual devices are defined differently, depending on which form of energy is converted into which other (see Eq. 6.1). We have to pay for the primary energy supplied, unless it is solar energy and its consequential effects. We receive the useful energy discharged as an equivalent value. From this point of view, anything with an efficiency of less than 50% seems rather unattractive. The efforts of the discoverers and inventors, the engineers and technicians, should therefore be directed towards overcoming this 50% limit. Examples of this are the development of the gas-and-steam power plant, power-heat-coupling, or the replacement of incandescent bulbs ($\eta \approx 7\%$) by LED lamps ($\eta \approx 50\%$) (for more details, see [11]).

We have already noted several times that the available solar energy could supply many thousands of times the needs of mankind. The various manifestations of energy contribute to this to quite different degrees (see Fig. 6.1). Table 6.2 shows what is possible from an *economic and ecological point of view*. These possibilities, the so-called **potentials,** are based on estimates which take into account the environment and the reservations of the population. Such figures should therefore only be viewed in terms of order of magnitude. They should be compared with the total final energy demand of mankind of 17,000 or 25,000 gigawatts in 2050 (see Fig. 4.4). In contrast, the potential of direct solar energy is very large. On land (excluding Antarctica), at least 230,000 gigawatts are available on average

Table 6.2 Ecologically and economically estimated usable potentials of alternative energy sources

Energy source	Energy form	Possible total power (GW)
Solar thermal and photovoltaics	Direct solar radiation	230,000 to 1 mill.
Wind		17,000 to 35,000
Rivers		1800
Biomass		8500
Tide	Gravitational energy	60
Geothermal energy	Radioactivity	70
Nuclear fission	Nuclear energy	500

worldwide. This is the 168 watts per square metre from Fig. 2.3, multiplied by one hundredth of the total continental area excluding Antarctica of 136 million km^2. So, even from an ecological and economic point of view, we have more than ten times the extrapolated actual demand available. Both wind and photovoltaics could provide much more energy than we need, mind you of the total energy. Electrical energy alone accounts for only about a quarter of that.

6.9 The Space Requirement [24]

One often hears the argument that the extraction of alternative energies would disfigure our environment and consume land that would be needed elsewhere, for example for agriculture. These fears are wrong, as we will now show. What makes our environment ugly is, of course, a matter for differing opinions. A "forest" of wind turbines is certainly something unfamiliar. But one will get used to it just as one gets used to our smokestacks, high-voltage and cell phone towers, or our freeways and highways with their forest of traffic signs. On the other hand, the amount of land needed for wind turbines and solar panels is minimal compared to our traffic and commercial areas. Indeed, we have enough unused land for solar energy generation. Let's explain:

In 2050, taking into account the necessary social redistribution, each person in industrialized countries should need an average of 4 kW of total power, and 2 kW in developing and newly industrializing countries (see Fig. 4.4). This results in a global average of 2.5 kW per capita, and we can assume that 25% of the population will then live in industrialized countries and 75% in the others. We now calculate how much area is needed if the energy required for this is supplied either only by solar cells or only by wind turbines:

A current solar cell module with a surface area of 1 m^2 delivers 100 W of electrical power at 1000 W irradiation (see Sect. 6.3). However, on average only 168 W of radiation per square metre are available worldwide (see Fig. 2.3), which means that the cells only supply 17 W per square metre in real terms. Dividing the necessary 2.5 kW per capita by the 17 W per square metre results in an area of 147 m^2 per person for the total energy demand. For the electricity demand alone, only a quarter of this would be necessary, namely 37 m^2 of solar cell area.

If, on the other hand, we were to cover all our needs with wind turbines, it should be borne in mind that they would not be allowed to stand arbitrarily close to each other. Otherwise they would "take each other's air". An optimal density is four wind turbines per square kilometer [24]. Because the wind is not always blowing, the maximum output of 5 MW per wind turbine is only about one third of this (see Sect. 6.7). This results in 6.8 MW per square kilometre. Dividing the demand of 2.5 kW per capita by this power density again results in a land requirement of 370 m^2 per person, and for electricity alone 93 m^2.

Now, with these numbers, consider that while solar cells need all of their land, wind turbines do not. Because you can have industry, commerce and agriculture under them. And you can even live under them if they are not too noisy. And the noise of the wind turbines is a matter of aerodynamics, which is being worked on. What these numbers mean can be seen in the following example: for the city of Stuttgart with 650,000 inhabitants, you would need about 1500 wind turbines for total energy, or 380 for electricity supply alone. Or you would need about 150 km^2 of solar panels or 38 for electricity alone. Of course, these areas do not have to be in Stuttgart itself, but in the North Sea or the Mediterranean for the wind turbines and in Spain or the Sahara for the solar cells. But then you would need about 10% more power for the transport losses. If all available roofs in Germany were covered with solar cells, we could cover our entire electricity demand [35].

Such numbers seem a bit daunting at first. It becomes quite different when you consider that most of the world's population does not live in large cities like Stuttgart, with its density of 6500 people per square kilometre. Worldwide, there will be an average of about 80 people per square kilometer in 2050. And these have 136 million km^2 of land at their disposal. Fig. 6.6 shows how this area is used on average. Among other things, each person has an average of 4500 m^2 of unused land. If one were to use the marked parts of this land for solar cells or wind turbines, this would "hurt nobody". For electricity alone, only a quarter of it would be needed. So we really do have enough space under the sun to satisfy all our energy needs!

6.10 Storing Energy [11]

The use of solar energy has a certain disadvantage compared to conventional energy supply, which must not be concealed: The sun does not always shine and the wind does not blow permanently or with the same strength. Both cannot simply be switched on and off as required. The power supply from solar cells and wind turbines therefore fluctuates over time. Demand also fluctuates, but usually not in the same sense (Fig. 6.7). However, there is in principle a very simple remedy for the fluctuations in the supply of solar energy: store the energy when you produce too much, and empty the store again when you need more. Heat and electricity can be stored in various ways. And what do such storage facilities look like?

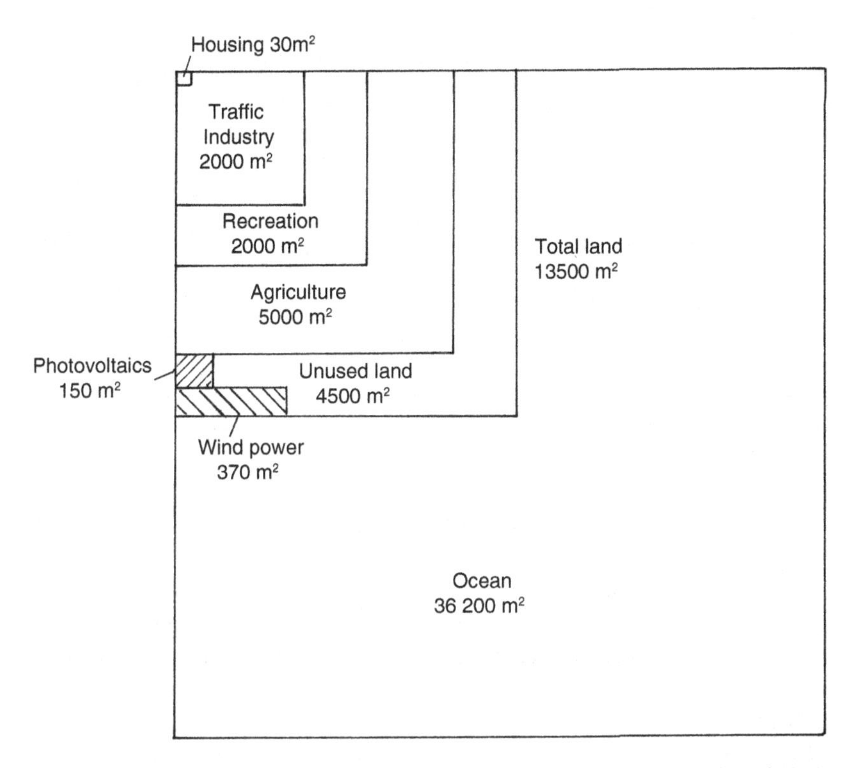

Fig. 6.6 Available land (excluding Antarctica) per person for a world population of 10 billion people with a Central European standard of living

The main storage methods for electrical energy are outlined in Fig. 6.8: Pressurised storage, accumulator batteries and chemical storage.

- In pressurized storage, either water is pumped into a reservoir at a higher elevation, and it delivers its potential energy as it flows down a turbine. Or air is pumped at high pressure into underground cavities, which is then allowed to flow out again through a turbine.
- How an accumulator works for storing electricity is generally known.
- In a chemical storage method, for example, hydrogen is produced by electrolysis and later converted back into heat or electrical energy by oxidation, for example in a fuel cell. The hydrogen can also be converted into methane (CH_4), which is more convenient to handle. In the process, CO_2 is consumed at the same time:

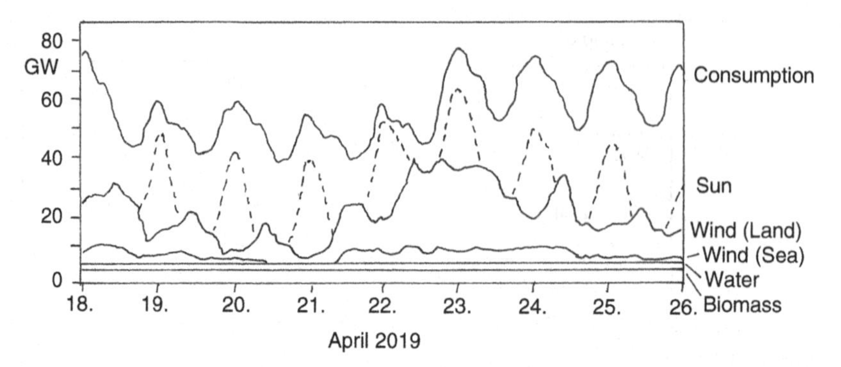

Fig. 6.7 Electricity consumption in Germany for one week in April 2019 and shares of the various alternative energies in it (After [32]). The period covered here was particularly rich in wind and sunshine

$$CO_2 + 4H_2 + 165\,kJ/mol \rightarrow CH_4 + 2H_2O. \tag{6.3}$$

The investment costs for these three methods are very different. Per kilowatt hour it is about 70 euros for a pumped storage plant, 100 euros for accumulators and 10,000 euros for hydrogen production. The latter is the most expensive method, but the product is also very versatile.

If you want to store heat rather than electrical energy, there are three main methods: temperature storage, latent heat storage and thermochemical storage. Either a larger delimited quantity of water or rock is heated and later cooled down again, for example with heat pumps. Or salts are melted and later solidified again. Or endothermic chemical reactions are started and the products are later transformed back exothermically. In all these cases, larger amounts of thermal energy can be conserved for a certain time and then recovered.

There is therefore a whole range of energy storage methods that can be used to compensate for the temporal fluctuations in their generation rate. Unfortunately, however, there are still far too few such facilities in existence worldwide. The construction of energy storage systems does not seem to be economically interesting so far. Such investments will only be worthwhile when conventional power plants, which can be switched on or off as required, are gradually decommissioned. This is also foreseeable due to climate change and the depletion of fossil fuels. The world's largest pumped storage plant with a capacity of 3 gigawatts is located in Virginia (USA), the largest German one with 0.37 gigawatts at Schluchsee in the Black Forest.

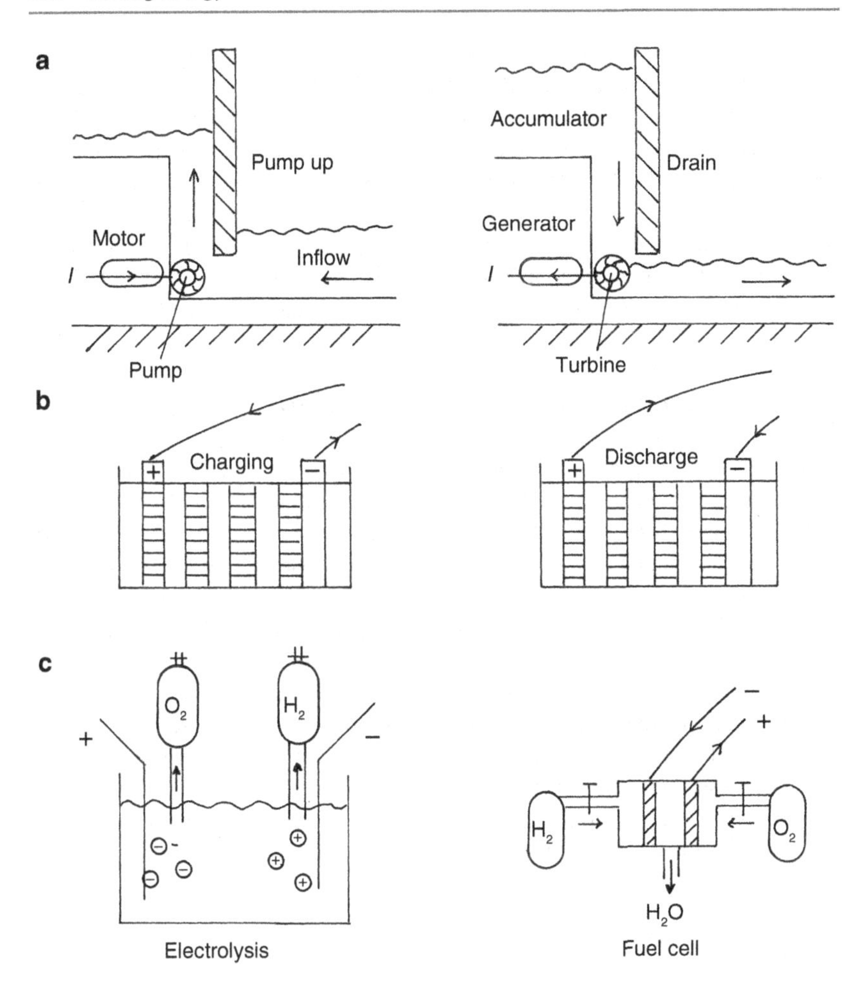

Fig. 6.8 Possibilities for storing electrical energy (storing on the left, discharging on the right). (**a**) Pumped storage, (**b**) Accumulator battery, (**c**) Water electrolysis

Germany currently has 45 GW of photovoltaic capacity installed and 58 GW of wind power. Together, these provide an average of 19 GW of electrical power, i.e. 27% of our total electricity demand of 70 GW. However, due to fluctuating weather, they only provide 20% of their maximum capacity (Federal Ministry for Economic Affairs and Energy 2019). In order to balance the fluctuations of solar and wind energy (see Fig. 6.7) in Germany, one would need storage facilities in the ten-gigawatt range with operating times of 1–2 weeks. This would be at least ten times as much as is available today, namely only 7 GW for about 6 h. It can be estimated that we should be able to store a maximum of 50 GW for 2 weeks to ensure full compensation for weather fluctuations. However, the storage problem is not quite as serious as it looks. Indeed, when the sun shines strongly, there is often little wind, and when the sky is overcast, there is more (see Fig. 6.7). Nature itself helps us to balance things out. If enough solar radiation and wind energy plants were operated worldwide, a large interconnected grid would bring about the balance autonomously. A small part of this has already been realized in Europe today.

6.11 Summary of the Energy Conversion

In this chapter, we discussed the most important energy converters that can convert solar radiation into electricity. These methods are now so far developed that they can be used worldwide. Nevertheless, we obtain only 2% of our energy needs from the sun, even though it could supply us with many thousands of times that amount! According to the forecasts of the world climate conferences, most or even all fossil fuels should be replaced by solar energy by the year 2050 in order to stabilize our climate at least to some extent. Then the threat of mass migrations from the flood plains could also be largely avoided. In addition, we would have saved the last reserves of coal, oil and natural gas for our descendants, because these raw materials are too valuable to burn them completely. However, it requires a consistent rethinking of our consumption habits and drastic economic policy measures on the part of governments to achieve these goals. Whether we are really capable of doing so, only the future will show [15].

7 Non-solar Energy Converters

7.1 Tidal Energy

The tides, the ebb and flow, are huge movements of water in the Earth's oceans. They come about through the attractive forces of the moon and the sun, as we learned in school. The tidal range is about 50 cm in the open ocean, but can reach 17 m on some coasts. There are two ways to harness its potential energy: Either you install turbines or propellers in the ocean, similar to river power plants, and let the tidal current drive them. Or one fills a dam at high water and balances its level again at low water by means of a turbine. One of the largest such power plants is located in the Rance river on the French Atlantic coast. It has a 750 m long dam wall and a peak electrical output of 240 megawatts. The difference in height between high and low water here is between 12 and 16 m. Such a power plant has an efficiency of up to 90%. There are about 100 favourable locations worldwide for such tidal power plants, which together could supply up to 100 gigawatts. However, this is only about half a percent of our total current energy demand (see Fig. 4.1).

7.2 Geothermal Energy

Our earth is very hot inside, 99% of its volume has a temperature of more than 1000°C, in the center even 7000°C. This heat essentially comes from the radioactive decay of the elements uranium, thorium and potassium in the earth's interior. The radiation released in the process heats up its surroundings and causes the heat to radiate through the Earth's surface into space at a rate of about 0.1 watts per square metre. This is not much compared to the 170 watts per square metre of solar

K. Stierstadt, *Our Climate and the Energy Problem*, essentials,
https://doi.org/10.1007/978-3-658-38313-8_7

radiation on this surface (see Fig. 2.3). The temperature of the Earth's solid crust increases with depth from the surface, by about 2.5° per 100 m. These temperature differences can be used with a heat-power machine. Figure 7.1 shows the principle of such a power plant. Hot water is transported from the depths by its natural pressure or by a pump to the surface. There it drives a liquid or gas circuit for a turbine and a generator. The water cools in the process and is pumped back into the ground to replenish the reservoir there. Worldwide, geothermal energy is currently used to generate about 20 gigawatts of electrical power, a few tens of megawatts per plant. This is only worthwhile where there is enough hot water at a depth that is not too great. In Germany, there are 25 experimental plants that extract water at a temperature of about 150°C from depths of up to 4 km.

7.3 Nuclear Energy [18, 19, 20]

A not entirely small part of our primary energy is still generated in Germany in nuclear reactors, namely about 5% (see Fig. 4.1). In the reactors, uranium atomic nuclei are split, and in return they supply heat and electricity. Although this does not produce carbon dioxide, the process is problematic for other reasons. The operation of the reactors produces waste products of uranium nucleus fission, many

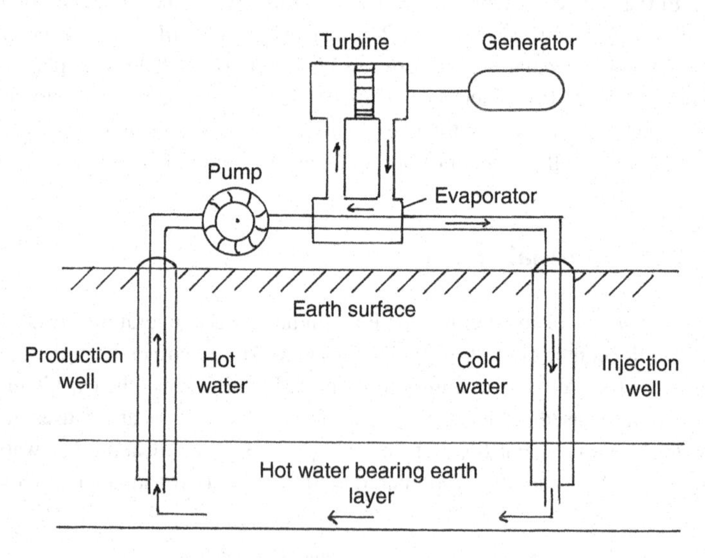

Fig. 7.1 Principle of a geothermal system

of which are life-threatening, **fission products** and **transuranic elements.** We learned how this happens at school. In a power plant reactor there are about 100 tons of uranium, arranged in 200 fuel elements, each 5 m long and 50 cm in diameter. During nuclear fission, heat is produced and the fuel elements become about 500°C hot. This heat is fed through a water circuit to a turbine which drives a generator. The fuel elements are used up after about 3 years and have to be disposed of. Their radioactive radiation is then so strong that one receives a lethal dose of radiation during only 10 s (!) at a distance of one meter from a spent fuel element. One then dies within a few days ("acute radiation death"). This radioactive radiation resembles to a large extent the X-rays, only it has a much higher energy. It kills the cells of all organic living beings by ionizing and destroying their molecules.

The waste from the reactors must therefore be handled with extreme care. After all, you can't get too close to them, and you can only handle them remotely. It is true that their radioactivity decreases over time, after a 100 years to about one thousandth of the initial value. But even then you still have to stay far away of such a fuel element, because you would get the lethal dose of radiation within 30 min at a distance of one meter. In addition, there is the high temperature of the spent fuel elements. It takes about 10 years for the temperature to drop from 500°C to 100°C. This takes such a long time because the radioactive elements have to decay. This takes so long because the radioactive radiation itself heats up all the matter in which it is produced. The fuel elements must therefore be cooled for 10 years in "decay pools" with water, and then for another 50 years in "interim storage facilities" with air.

Today, there are about 500 power plant reactors in operation worldwide. In Germany, due to the "nuclear phase-out", there are currently only 6 in 3 nuclear power plants. The last one is scheduled to be shut down in 2022. Many hundreds of thousands of fuel elements from the past 60 years of nuclear energy use are now in above-ground storage halls and are "radiating away", in Germany alone about 50,000. Their radioactivity poses a constant latent danger. An accident, such as the most recent one in Fukushima (Japan) in 2011, in which several thousand fuel elements were damaged, contaminates tens of thousands of square kilometres of land, rendering it uninhabitable for decades. In Fukushima, an area of 30 km radius with 120,000 people was completely evacuated.

How can we now cope with this legacy that the reactor operators have left us? Since radioactivity cannot be influenced or even prevented by any known method, one must try to live with it – but at the greatest possible distance. The best way is to put the waste deep in the ground. Then the radiation can no longer be dangerous to us, because it is absorbed by sufficiently thick layers of matter, i.e. it is converted into heat. One metre of sand or rock attenuates it to about one hundredth. When dumping in the ground, however, it is important to note that the radioactive substances must not come into contact with groundwater and dissolve in it, which would contaminate it with radioactivity. Therefore, the spent fuel elements are placed several hundred meters deep into the earth in watertight and corrosion-resistant containers. There they are to remain safely for at least a hundred thousand years until their activity has sufficiently subsided. What might happen to the waste in a hundred thousand years on earth and below, however, is another question.

Unfortunately, all this is very expensive, because such a waste repository must be protected against water ingress and against accidental destruction by rock shifting, earthquakes, volcanic activity and unauthorized access. For these reasons, there is not yet a single such repository for nuclear power plant waste, or **nuclear waste,** anywhere in the world. Only in the USA is there a small underground storage facility for military waste. It is true that about ten percent of the profits of the nuclear energy companies would have been enough to dispose of all the waste accumulated so far in this way. But this was not pushed through politically for a variety of reasons: It would have increased the price of electricity by perhaps ten percent. And the resistance of the population to such storage facilities would have had to be overcome by proper information. This has not happened so far. Only in Finland has there been any success: such a repository is under construction near Olkiluoto on the Baltic Gulf, about 400 m deep in granite rock. Here, the waste from the four Finnish nuclear power plants is to be safely deposited for 100,000 years. Germany has been planning to build such a repository for 30 years (e.g. Gorleben), which according to various estimates would cost 20–60 billion euros, but so far without any results. As long as the nuclear waste remains above ground in makeshift storage facilities, it is not protected, neither against natural disasters (see Fukushima), terrorism, acts of war (e.g. Ukraine), power failures, airplane crashes, etc. The nuclear waste problem is thus one of the great challenges of our time, along with climate change, population growth, and perhaps cancer control. By way of illustration, Fig. 7.2 shows an overview of the uranium cycle and a realistic disposal concept (for more details, see [18]).

7.4 Breeder Reactor [11]

The known stocks of uranium fuel for nuclear reactors will run out in 100–150 years at current consumption rates (see Fig. 4.5). But consumption will continue to rise (see Fig. 4.4). We are therefore considering what could take the place of uranium fission thereafter, if nuclear energy is still needed at all. A well-known fact comes to the rescue: in conventional nuclear reactors, the uranium atomic nuclei are split

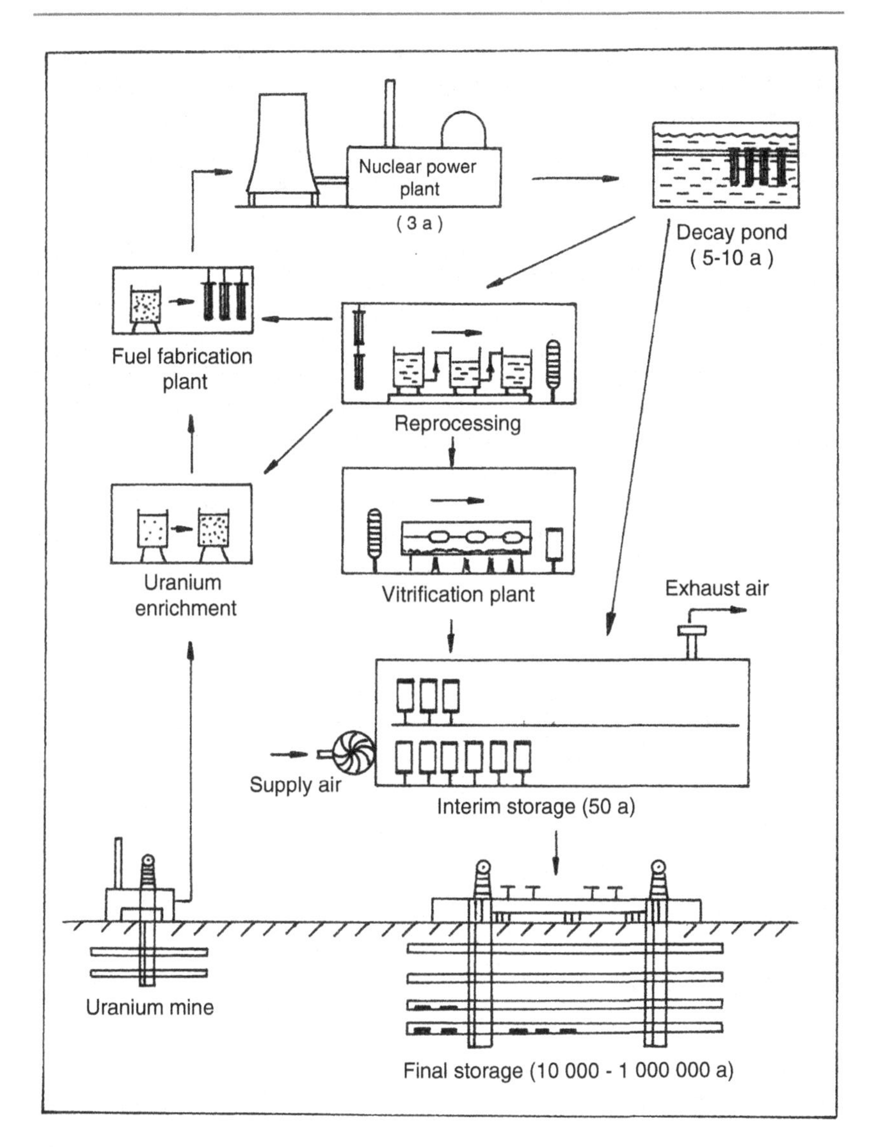

Fig. 7.2 A realistic concept for nuclear waste disposal (a: years)

by neutrons. And many of these fission processes produce two to three new neutrons. That is slightly more than is needed to operate the reactor and replace its fuel at the same time, namely only two new neutrons. The surplus third can then be used to convert other atomic nuclei into those that can also be split with neutrons, such as the uranium-235[1] currently used. Two candidates are available for this method: the isotopes uranium-238 and thorium-232. If their atomic nuclei are bombarded with fast neutrons, the isotopes plutonium-239 and uranium-233 are produced, which are suitable as fuel. A reactor can be operated with both of them, similar to the uranium-235 used so far.

In order to produce the new fuels, the core of a conventional reactor is surrounded by a jacket, the "breeding material", made of uranium-238 or thorium-232 (Fig. 7.3). The excess neutrons of uranium-235 then convert uranium-238 into plutonium-239 or thorium-232 into uranium-233, in each case by means of two beta decays. In this way, the new fuel has been obtained, about 1.3 times more than was consumed in the core of the breeder reactor. This means that the old fuel has been replaced and at the same time 30% of new fuel has been produced.

Unfortunately, as is often the case, there is a catch: a breeder reactor is a highly dangerous machine. It works with fast neutrons instead of slow ones like a conventional one and is therefore also called a "fast breeder". It therefore gets 900°C hot and has to be cooled with liquid sodium at 10 bar pressure instead of water. The

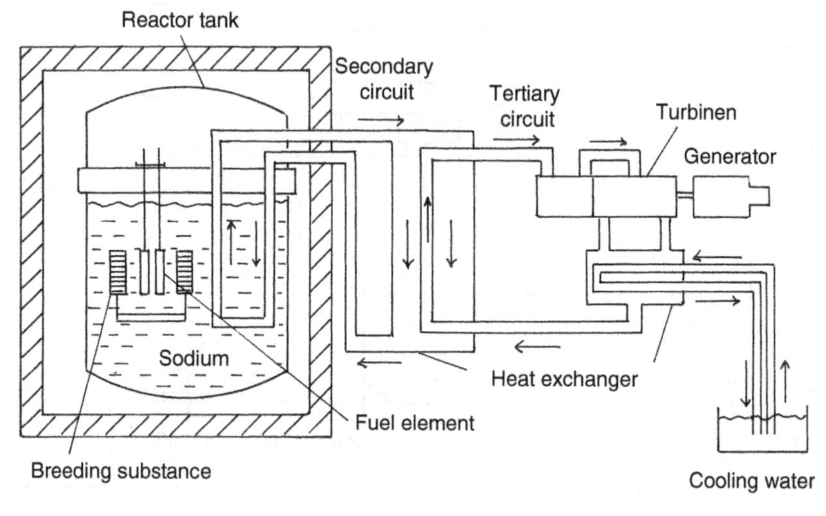

Fig. 7.3 Principle of a breeder reactor

[1] The *mass number* attached to the element name denotes the number of nucleons (protons and neutrons) in the atomic nucleus.

water would slow down and absorb the valuable excess neutrons. However, hot liquid sodium is a very unpleasant substance: it attacks many metals and ignites by itself when exposed to air. The associated technical problems cannot yet be fully controlled. Therefore, of the 22 breeder reactors built to date, only 2 remain [21]. All others were either destroyed during operation or shut down as a precaution. So for the time being, the nuclear fuel breeder process remains a future dream. In addition, there is the nuclear waste problem from Sect. 7.3. A breeder reactor produces even more radioactive waste than a conventional one. And there is as yet no safe way of disposing of the waste. Moreover, for the reasons described above, operating a breeder reactor is much more complex and expensive than operating a conventional one. The electricity generated with the breeder products would therefore also become considerably more expensive.

7.5 Nuclear Fusion [22]

Now we come to an even more uncertain or unrealistic method of non-solar energy production, **controlled nuclear** fusion (keyword: "artificial sun"). If hydrogen atomic nuclei are allowed to collide with each other at a temperature of 100 million degrees (!), then helium atomic nuclei are produced from them, and a great deal of bonding energy is released in the form of heat, namely 120,000 kWh per gram of helium! That would be enough to supply energy to 1000 people in Germany for a whole day. This helium fusion is the same process that takes place in our sun and provides its enormous radiation power (see Fig. 5.3). The only difference is that in the sun, because of its great density, this process "already" takes place at a temperature of 10 million degrees.

So you want to "imitate the sun". But it's not so easy to generate 100 million degrees on Earth. In 2011, work began on building a huge fusion reactor called ITER (International Thermonuclear Experimental Reactor) at the French research centre of Caderache, 60 km north of Marseille. This is one of mankind's biggest scientific projects, along with the manned moon landing, the International Space Station and the CERN nuclear research centre in Geneva. The European Union, China, India, Japan, Russia, South Korea and the USA are participating in the construction and financing of ITER. The reactor will be about 30 m high and have the same diameter (Fig. 7.4). Inside it is a torus-shaped fusion vessel in which the

Fig. 7.4 Construction of the ITER fusion reactor. The solenoid for accelerating the plasma and the toroidal fusion vessel with a kidney-shaped cross-section can be seen in the middle (for size comparison: person on the bottom right). (Photo: ITER Organization)

fusion of the heavy hydrogen nuclei deuterium (H-2 or D) and tritium (H-3 or T) into helium (He-4) is to take place[2]:

$$D + T \rightarrow He{-}4 + 17.6\,\text{MeV}. \tag{7.1}$$

With a heating power of 50,000 kW, an output power of 500,000 kW is to be achieved, i.e. a nominal efficiency of 10 or 1000%. Whether this will succeed is still uncertain for the following reason: The deuterium-tritium plasma is concentrated and accelerated in the ring centre of the torus by huge magnetic fields of up to 12 Tesla strength. The superconducting coils for these fields can be seen in the figure. It can happen that the plasma stream becomes unstable and touches the wall

[2] The reaction energy of 17.6 MeV corresponds to $2.8 * 10^{-12}$ (S.M.) joules per helium atom or 1.7 trillion joules per mole.

of the torus. In that case, fusion ceases immediately. And these instabilities cannot be calculated and predicted exactly. It is also not yet entirely clear whether the dissipation of the fusion heat through certain wall areas of the torus is successful, and what proportion of the energy escapes as electromagnetic radiation. The success of the experiment is therefore uncertain, and for this reason the USA has temporarily left the consortium.

Nevertheless, large sums are being invested in the project. According to the latest forecasts, it will cost at least 15 billion euros (originally planned at 5 billion) and will not be ready for initial tests until 2028 at the earliest. The decisive fusion experiment is not envisaged until 2035. This gigantic project resembles, as said, the International Space Station (ISS), only its outcome is questionable, while the space station has been functioning for years. Whether nuclear fusion can ever be used to generate electricity at a reasonable cost is anyone's guess. Our natural sun is certainly much, much cheaper.

Herewith we have learned about the non-solar energy sources discussed today, all of which do not produce carbon dioxide. However, it does not look as if any of them will be able to make a remarkable contribution to our energy supply in the foreseeable future. And therefore there is still no prospect of being able to stabilise our climate in this way.

Epilogue: Afterword – How Can We Go On?

We have now discussed our energy problem, the cause of climate change, at length. We have discussed the demand and sources of energy and its methods of conversion into motion, electricity and heat. Now we come back to the beginning of our considerations, to the climate and its development.

Our climate is changing dramatically. Temperature and sea level are rising worldwide. The causes for this are based on human activity. Even the climate optimists (see preface) can no longer explain this away. The foreseeable consequences of this development are extremely unpleasant or catastrophic for mankind as a whole. The United Nations recognized this early on. In 1988 they founded the "International Commission on Climate Change" (IPCC, International Panel on Climate Change). This consists of several hundred experts from 90 countries, who register and evaluate all observations and climate changes and publish the forecasts [23].

In addition to this UN Commission, there are two series of conferences that take place on behalf of the United Nations at annual intervals: the "World Climate Conference" and the "World Population Conference". Unfortunately, these conferences, like the Climate Commission, can only make recommendations, but cannot pass resolutions or order concrete measures. Therefore, despite decades of warnings, there is still no hope of real improvement for our climate. The temperature on the earth's surface continues to rise every year and the sea level continues to rise at an accelerated rate. The reasons for this are primarily of an economic nature. We would have more than enough climate-neutral solar energy, but we still use it far too little. This is because today's conventional and climate-damaging energy

K. Stierstadt, *Our Climate and the Energy Problem*, essentials,
https://doi.org/10.1007/978-3-658-38313-8

technology with its carbon dioxide production brings large profits to the economy, while clean solar technology still requires large investments. The only way out of this dilemma would be a consistent rethink, as well as changes in our consumer behaviour and drastic economic policy measures. But as long as the fires are only raging in Australia and the floods are only threatening the Pacific islands and Bangladesh, nothing decisive is happening in Europe, China or North America to save the climate. The St. Florian principle applies here: "Oh holy St. Florian, spare my house, set fire to others".

Thus, from today's perspective, we have only two possible alternatives for our future behavior [5]:

- Either continue as before. That means burning all known fossil raw material reserves bit by bit. Then the temperature of our biosphere will rise by about 6.5°–9.5° in 100–200 years. That must result in global shifts in the foundations of our food supply. And sea levels will rise by 60 m in the next 400 years, namely when the Antarctic ice will have melted completely. This will flood large parts of the inhabited mainland.
- If, on the other hand, we manage to halve the burning of coal, oil and gas worldwide by about 2035 and to stop it completely by 2050, then the following will happen: the temperature of the biosphere will rise by no more than 2–3° and the sea level by 1–2 m by about 2100. This will result in some losses of inhabited land, but perhaps not too drastic for our food economy.

The fact that the second alternative is also politically feasible and, above all, affordable is shown in the following Fig. A1. The producer price for electricity from solar energy is already just as low today as that from burning lignite, namely 6 cents per kilowatt hour on average. And the investments required for a complete conversion of power generation to solar technology are certainly considerably lower than those required for the promotion of nuclear energy technology in the second half of the last century. Those were on the order of half a trillion dollars worldwide. Installing solar worldwide should certainly be considerably cheaper. And we could do it.

What You Can Take Away from This *Essential*

- The current global warming and sea level rise are man-made and based on the burning of coal, oil and natural gas.
- If this practice continues, the temperature of the biosphere will continue to rise and sea levels will continue to rise threateningly.

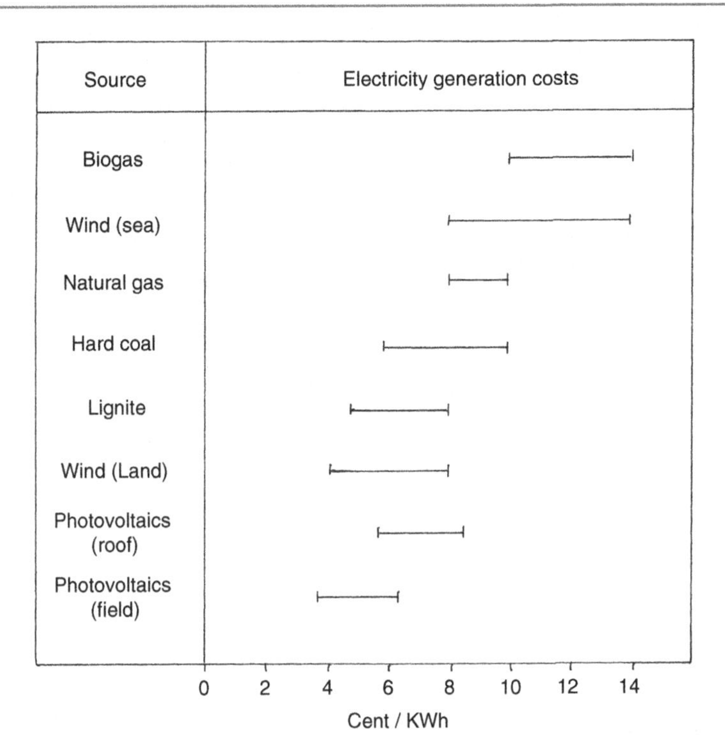

Fig. A1 Electricity production costs in cents per kilowatt hour in Germany 2018. (According to: www.Fraunhofer-Institut.ISE)

- Instead, solar energy could cover most of our needs. Economic habits of thought and behaviour stand in the way of this. Too much is earned with fossil fuels.
- They learned how the sun works and what devices can be used to convert its energy into electricity.
- For the solution of the energy problem we have two alternatives: *Either* continue to burn coal, oil and gas. Then the temperature in the biosphere will rise by up to 10° and the sea level by up to 60 m: There is a "man-made warm period". *Or* replacement of fossil fuels by solar energy within the next 30 years. Then the temperature will rise by only about 2° and the sea level by about one meter.

Literature

Feldmann, D. R., u. a. (2015). *Observational determination of surface radiative forcing by CO_2 from 2000 to 2010*. Nature **519**, 339.
N. N. (2019). *Idealisiertes Treibhausmodell*. https://de.wikipedia.org/wiki/.
Karl, T. R. u. Trenberth, K. R. (2003). *Modern Global Climate Change*. Science **302**, 1719.
N. N. (2020). *Klimafakten*. https://www.dwd.de/Klima.
N. N. (2019). *Globale Erwärmung*.https://de.wikipedia.org/wiki/.
Wolfson, R. (2008). *Energy, Environment, and Climate*. New York: Norton & Co.
N. N. (2020). *Meeresspiegelanstieg seit 1850*. https://de.wikipedia.org/wiki/.
N. N. (2019). *Kohlenstoffdioxid*. https://de.wikipedia.org/wiki/.
Bundesministerium für Wirtschaft und Energie (2019). *Energiedaten*. Berlin: BMWi.
Umweltbundesamt (2020). (*Schriftenreihe*). www.Umweltbundesamt.
Stierstadt, K. (2015). *Energie – Das Problem und die Wende*. Haan-Gruiten: Europa-Lehrmittel.
Taube, M. (1985). *Evolution of Matter and Energy*. New York: Springer.
Zeilik, M. u. Gaustad, J. (1983). *Astronomy*. Cambridge (USA): Harper & Row.
Goetzberger, A. u. a. (1997). *Sonnenenergie: Photovoltaik*. Stuttgart: Teubner.
Stierstadt, K. (1995). *Sind wir zu dumm zum Weiterleben?* Salzburg: Annalen der Europäischen Akademie für Wissenschaften und Künste, Bd. 10.
Stierstadt, K. (2019). *Thermodynamik für das Bachelorstudium*. Berlin: Springer.
Bundesministerium für Wirtschaft und Energie (2019). *Erneuerbare Energie*. Berlin: BMWi.
Stierstadt, K. (2010). *Atommüll – wohin damit?* Frankfurt/M: Harri Deutsch.
Krieger, H. (2002). *Strahlenphysik, Dosimetrie und Strahlenschutz*. Stuttgart:Teubner.
De Santis, E., Monti, S. u. Ripani, M. (2016). *Energy from Nuclear Fission*. Swizerland: Springer.
N. N. (2020). *Brutreaktor*. https://de.wikipedia.org/wiki/.
N. N. (2020). *ITER*. https://de.wikipedia.org/wiki/.
N. N. (2000–2018). *IPCC-Berichte*. https://www.ipcc.ch/assessment-report.

K. Stierstadt, *Our Climate and the Energy Problem*, essentials,
https://doi.org/10.1007/978-3-658-38313-8

Stierstadt, K. (2019). *Genug Platz an der Sonne.* Physik in unserer Zeit **50**, 128.

N. N. (2019). *Treibhauseffekt.* https://de.wikipedia.org/wiki/.

Voosen, P. (2018). *The Realist.* Science **359**, 1320.

Moberg, A. u. a. (2005). *Highly variable Northern Hemisphere temperatures reconstructed from low- and high-resolution proxy data.* Nature **433**, 613.

N. N. (2020). *Mittelalterliche Warmzeit.* https://wiki.bildungsserver.de/klimawandel/index.php/.

N. N. (2020). *Mittelalterliche Klimaanomalie.* https://de.wikipedia.org/wiki/.

Petit, J. R. u. a. (1999). *Climate and atmospheric history of the past 400,000 years from the Vostok ice core, Antarctica.* Macmillan Magazine **399**, 429.

Falkowski, P. u. a. (2000). *The Global Carbon Cycle: A Test of Our Knowledge of Earth as a System.* Science **290,** 291.

Graichen, P. u. a. (2020). *Die Energiewende im Stromsektor: Stand der Dinge 2019.* Berlin: Agora Energiewende.

N. N. (2019). *Weltenergiebedarf.* https://de.wikipedia.org/wiki/.

N. N. (2020). *CO_2-Emissionen steigen weiter.* Phys. i. u. Zeit **51**, 11.

Umwelt-Bundesamt (2016). Siedlungs- und Verkehrsflächen & Struktur der Flächennutzung. Bericht, Dessau-Roßlau.

Printed in the United States
by Baker & Taylor Publisher Services